Graphical Models and Causal Discovery with R

Joe Suzuki

Graphical Models and Causal Discovery with R

100 Exercises for Building Logic

Joe Suzuki
Graduate School of Engineering Sciences
Osaka University
Toyonaka, Osaka, Japan

ISBN 978-981-95-4266-6 ISBN 978-981-95-4267-3 (eBook)
https://doi.org/10.1007/978-981-95-4267-3

This Springer imprint is published by the registered company Springer Nature Singapore Pte Ltd.
The registered company address is: 152 Beach Road, #21-01/04 Gateway East, Singapore 189721, Singapore

Preface

Graphical models provide a highly attractive framework that captures complex relationships among random variables both visually and mathematically. By fusing probability theory with graph theory, methods such as Bayesian networks and Markov networks have been systematized to model real-world structures. With these models, we can go beyond merely grasping correlations and move toward *causal discovery*, which reveals causal structures among variables.

Causal discovery, the topic of this book, consists of two key steps: not only estimating the pattern of dependencies (the "skeleton") among variables but also placing arrows to determine the *causal ordering* of variables. This problem had been difficult to treat within traditional statistical frameworks alone, but recent advances—kernel methods, the use of non-Gaussianity, information criteria, and Bayesian marginal likelihood—have enabled more flexible and theoretically coherent approaches.

One of the charms of graphical models is that they provide an intuitive sense of "structure" that cannot be obtained from formulas or code alone. You can draw them as figures, explain them in words, and manipulate them by hand to understand them—this visual and operational understanding brings the abstract world of probability and statistical modeling much closer. Standing at the intersection of statistics and machine learning, and bridging theory and practice, this field will undoubtedly play an increasingly important role.

My encounter with graphical models dates back to the early 1990s, before the term "machine learning" had become as common as it is today. While an assistant at Waseda University, I happened upon the proceedings of the UAI (Uncertainty in Artificial Intelligence) conference at an English-language technical bookstore along Meiji-dori near the School of Science and Engineering. I was immediately drawn into the refined discussions on how to learn the structure of Bayesian networks from data.

Around the same time, I was introduced to Jorma Rissanen's MDL (Minimum Description Length) principle and strongly resonated with the idea of "discovering the simple structure behind data." I promptly applied MDL to the problem of learning graph structures from samples and presented the results at UAI in 1993.

I have been told that this was among the earliest presentations by a Japanese researcher to introduce structural learning of graphical models to the international community.

Since then, I have been captivated by the potential of graphical models and have continued my research. Behind diagrams that may look deceptively simple at first glance lie a large number of mathematical challenges. Facing each of them and carefully untangling them has been the starting point of my career as a researcher and remains at the core of my work today.

This book tackles two major tasks in causal discovery: exploring dependency structures and exploring variable orders.

Causal discovery = exploring dependencies + exploring variable order

Given p variables and n samples, neither step alone suffices to draw the correct causal structure. If you simply feed your data into a tool, it will readily return something that looks like dependencies and an ordering. However, when the results differ from your expectations or run counter to intuition, unless you can explain *why* the algorithm behaved that way, you are left with mere outcomes. You will struggle to write a convincing paper or to persuade your supervisor or client.

This book focuses specifically on causal discovery within graphical models, incorporating modern techniques such as kernel methods and Bayesian approaches. For LiNGAM—the directed approach to exploring variable order devised by Professor Shohei Shimizu (Osaka University)—I have aimed to describe its essence clearly.

Furthermore, I have made extensive use of exercises from a graduate course taught in 2019 and materials from more than ten iterations of an online small-group seminar. As with my previous "100 Problems in the Mathematics of Machine Learning" series, the goal is not merely to memorize procedures but to think through "why it works," and to deliver a book that genuinely inspires.

The structure of this book is designed to deepen understanding step by step while balancing theory and implementation, intuition and rigor. The diagram below visually summarizes the relationships and roles of each chapter.

This is not a comprehensive text on all of graphical models; rather, it is a specialized book that delves deeply into the theory and practice of causal discovery. In particular, it has several distinctive features not commonly found elsewhere:

1. Instead of covering all of Bayesian networks, we specialize in the theory and methods of causal discovery. In particular, for LiNGAM and score-based structure learning, we go from theory down to implementation in detail.
2. For kernel-based statistical tests (HSIC, KCI, etc.), we concisely and intuitively describe their mathematical essence with an eye toward applications in the Peter–Clark (PC) algorithm and LiNGAM.

3. As Bayesian approaches to structure learning, we formulate marginal likelihood and its relationship to information criteria (especially BIC), and carefully unpack the underlying theory, including clear implications and limitations of BDeu and Jeffreys priors.
4. We do not stop at introducing concepts; wherever possible we include implementations in R and Python so that readers can learn in a reproducible manner.
5. We keep beginners in mind by explaining key ideas through diagrams and examples to build intuition. Even advanced topics are made more approachable through visual aids and a staged presentation.

While the primary focus is a practical understanding of causal discovery, for readers wishing to dig deeper into theoretical background, the appendices cover advanced theorems and derivations such as:

- The Darmois–Skitovitch theorem
- Murphy's derivation of marginal likelihood (normal mean with inverse-Wishart prior)
- The Verma–Pearl fundamental results on structure learning

These are intended for readers who wish to further enrich their mathematical understanding of causal discovery and are not mandatory for following the main text. Those uneasy with heavy mathematics can safely skip them; those who are curious will find solid, satisfying background explanations.

We summarize the flow of chapters as follows.

Chapter 1 reviews the basics of probability and statistics necessary for understanding the causal discovery methods developed in the second half of the book. Chapter 2 introduces the minimal concepts in graphical models such as conditional independence and separation. Chapter 3 treats kernel-based tests of independence and conditional independence (HSIC, KCI, etc.).

Chapter 4 introduces the PC algorithm that uses these tests. It is a constraint-based structure learning method that identifies the dependency skeleton of variables via conditional independence testing. Chapter 5 covers LiNGAM, which estimates causal order based on non-Gaussianity; in DirectLiNGAM, the independence tests of Chap. 3 play a key role.

Chapter 6 provides the theoretical background for score-based structure learning, covering information criteria (AIC, BIC) and the formulation and properties of marginal likelihood. Chapter 7 presents score-based structure learning methods that leverage these ideas. Like the PC algorithm, they estimate the variable structure, but do so by evaluating scores to compare candidate structures and select an optimal model.

In the figure, red boxes indicate chapters on foundational theory and blue boxes indicate chapters on concrete methods for causal discovery.

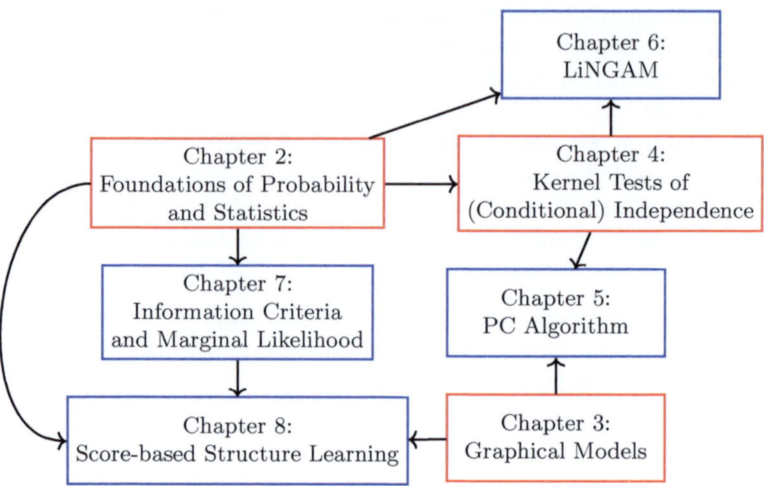

This book aims to face the challenge of causal discovery head-on while conveying to students, practitioners, and researchers alike both the "intellectual elegance" and the "practical power" of graphical models. I sincerely hope that all readers who pick up this volume will deepen their understanding of the rich world of causal discovery.

Features of This Series

Rather than this book alone, the distinctive features of the *series* are summarized below.

1. Make it your own: build the logic
   You grasp concepts mathematically, implement programs, and verify behavior by running them. By repeating this cycle, you will build the "logic" in your mind. You acquire not only knowledge of machine learning but also a way of looking at problems, enabling you to keep pace. Most students say, "I learned a lot," after solving the 100 problems.
2. Not just talk: code so you can move from code to action
   Books on machine learning without source code are inconvenient. Even if a package exists, without source code you cannot improve the algorithm. Sometimes code is on Git, but only for MATLAB or Python, or not sufficient. In this series, most procedures come with code; even if the math is difficult, you can understand what the code is doing.
3. Beyond "how to use": a scholarly text written by a professor
   Books full of package how-tos and examples can help beginners get started, but merely following steps without understanding the underlying operations is unsatisfying. Here, we present the mathematical principles with the code

that realizes them, leaving little room for doubt. This series leans toward the academic, full-fledged end of the spectrum.

4. Solve 100 problems: university exercises refined by student feedback
   The exercises have been used in seminars and lectures, iteratively improved through student feedback, and distilled into an optimal set of 100 problems. Each chapter's main text serves as the explanation; by reading it, you will be able to solve all exercises.
5. Self-contained within the book
   Have you ever felt let down by a proof that says "see Reference ○○ for details"? Unless highly motivated, few readers will track down such references. We choose topics to minimize the need for external citations. Proofs are kept as straightforward derivations; more difficult arguments are placed in appendices at the end of each chapter.
6. Not sell-and-forget: videos, online Q&A, and program files
   In university courses, we answer student questions on Slack around the clock. For this series, we provide a reader page

   https://bayesnet.org/books_jp

   to facilitate casual interaction between authors and readers. We also publish 10–15 minute videos for each chapter. Moreover, programs in the book are available for download from Git.

Toyonaka, Osaka, Japan Joe Suzuki

Contents

Chapter 1
A Gentle Introduction to Causal Discovery

In this chapter, as a warm-up, I would like to describe the overall picture of causal discovery without going into rigorous details. Using examples, I explain the two problems that make up causal discovery—dependency structure and variable ordering. My goal is for this single chapter to convey not only the big picture of the book but also its essence.

1.1 Causal Discovery = Discovering Dependencies + Discovering Variable Order

By representing the "follow/followed-by" relations on social media as directed edges, we can form a directed network (Fig. 1.1). Such visualizations are meaningful in that they reveal structures like central "hubs" and "isolated nodes" with no connections, and they can help us discover latent relationships and features.

In contrast, what this book deals with are networks that represent mathematically definable relationships such as probability and causality. In particular, we assume that each vertex corresponds to an actual measured variable (e.g., presence/absence of disease, smoking status). Based on their causal relations (dependencies, ordering), we consider whether to connect each pair of vertices with an edge, and if so, in which direction.

First, a graph that represents dependency structure with undirected edges is called a *Markov network* (MN), and a graph whose edges are all directed and form no cycles (a directed acyclic graph, DAG) is called a *Bayesian network* (BN). Among Bayesian networks, when we orient the arrows from causes to effects, we call it a *causal network* (CN).

Here, I will explain the distinction between Bayesian networks (which reflect dependencies) and causal networks (which additionally reflect causation) using the Asia dataset [15]. In Asia, we imagine a physician asking a patient whether

J. Suzuki, *Graphical Models and Causal Discovery with R*,
https://doi.org/10.1007/978-981-95-4267-3_1

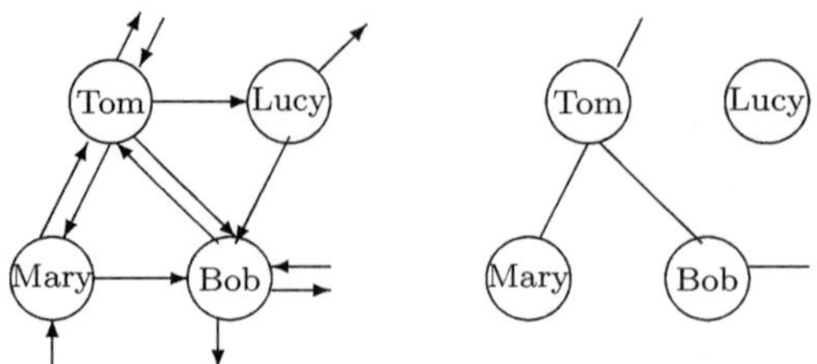

Fig. 1.1 A directed graph representing the relationship on Social Networking Service (SNS) where (following) →(being followed) (left), and an undirected graph showing only the mutual-follow relationships (right). Tom, Lucy, Mary, and Bob are the users

Table 1.1 Variables included in the Asia dataset

Variable	Meaning	Values
`Asia`	Recent travel to Asia	Yes/No
`Smoke`	Smoker	Yes/No
`Tuberculosis`	Tuberculosis	Yes/No
`Lung cancer`	Lung cancer	Yes/No
`Bronchitis`	Bronchitis	Yes/No
`Either`	Either tuberculosis or lung cancer	Yes/No
`X-ray`	Chest X-ray result positive	Positive/Negative
`Dyspnoea`	Shortness of breath	Yes/No

they recently traveled to Asia, whether they smoke, and so on, and diagnosing the likelihood of tuberculosis or lung cancer based on that information. Suppose, based on experience, the physician posits the causal relations below among eight random variables as listed in Table 1.1 and constructs the causal network shown in Fig. 1.2a. Throughout, " →" points from the cause on the left to the effect on the right.

Asia → *Tuberculosis* (Travel to Asia increases the risk of tuberculosis.)
Smoke → *Lung Cancer and Smoke* → *Bronchitis* (Smoking increases the risks of lung cancer and bronchitis.)
Tuberculosis → *Either and Lung Cancer* → *Either* (If either tuberculosis or lung cancer is present, *Either* becomes Yes.)
Either → *Xray* (When *Either* is Yes, an X-ray abnormality is more likely.)
Lung Cancer → *Dyspnoea and Bronchitis* → *Dyspnoea* (Lung cancer and bronchitis cause shortness of breath.)

The relations in Fig. 1.2 also state that the joint probability of the eight variables (denoted by initial letters) $P(A, S, T, L, B, E, X, D)$ can be written as a product of conditional probabilities:

$$P(X \mid E)P(E \mid T, L)P(T \mid A)P(A)P(L \mid S)P(B \mid S)P(S)P(D \mid L, B). \tag{1.1}$$

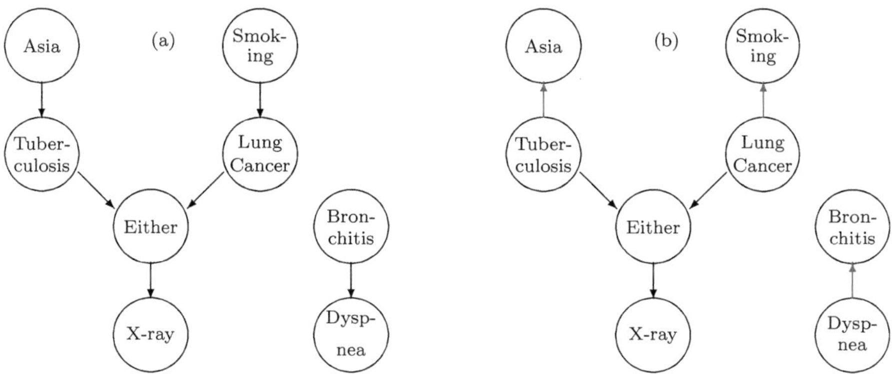

Fig. 1.2 The causal network derived from physicians' beliefs (**a**) and the Markov equivalent Bayesian network (**b**). In terms of dependencies, as seen from the fact that (1.1) and (1.2) are equal, the two are equivalent Bayesian networks. In the causal network, the arrow directions carry causal meaning, whereas in the Bayesian network, this is not necessarily the case

The aim of this book is to construct a network like Fig. 1.2a automatically from n independent and identically distributed (I.I.D.) samples on p variables (here, $p = 8$) without relying on domain experts. Sometimes such experts are not available; in other cases, forecasts based on past common sense may no longer apply. In such situations, discovering structure from data is effective.

For example, in structure learning of Bayesian networks studied in Chap. 8, one may conclude that the variable *Asia* is independent of the other seven variables. In the 1980s, travel from Europe or North America to Asia was relatively rare, and such a prior belief might have existed.

However, noting Bayes' theorem,

$$P(T \mid A)P(A) = P(T, A) = P(A \mid T)P(T),$$

we can rewrite (1.1) as

$$P(X \mid E)P(E \mid T, L)P(A \mid T)P(T)P(S \mid L)P(L)P(B \mid D)P(D). \tag{1.2}$$

Thus, Fig. 1.2a and b represent the same dependencies (we say they are *Markov equivalent*).

In general, a Bayesian network expresses dependencies among variables rather than causal order. It expresses a joint distribution as a product of conditionals and depicts that factorization with a DAG. Therefore, for each variable ordering, there is a corresponding Bayesian network.

Fig. 1.3 Bayesian networks in which the random variables X, Y, Z are connected in this order. (**a**), (**b**), and (**c**) represent the same dependency structure. (**d**) is called a collider, in which case X and Z become independent

For instance, consider three variables X, Y, Z connected in that order as a Bayesian network as in Fig. 1.3. In (a),

$$P(X \mid Y)P(Y \mid Z)P(Z) = P(X \mid Y)P(Y)P(Z \mid Y) = P(X)P(Y \mid X)P(Z \mid Y) = \frac{P(X, Y)P(Y, Z)}{P(Y)},$$

so it represents the same distribution as (b) and (c). By contrast, the distribution in (d),

$$P(X)P(Y \mid X, Z)P(Z) = \frac{P(X)P(X, Y, Z)P(Z)}{P(X, Z)},$$

differs. At this point, one might object: "Then why do Bayesian networks have arrows at all?" I will clarify this in the latter half of the chapter.

In our running example, Fig. 1.2b is also a valid Bayesian network. From data alone, we can conclude only that the truth lies in one of the Markov-equivalent factorizations like (1.1) or (1.2); in Fig. 1.2b, there are three red directed edges whose directions can be flipped independently, so there are $2^3 = 8$ Markov-equivalent Bayesian networks.

In general, standard statistical knowledge alone cannot deduce variable order. If, however, the causal order of variables is known, then among the Markov-equivalent networks, at most one is compatible with that order, so we need only search for the dependencies. For example, if the order

$$A \to T \to S \to L \to E \to X \to B \to D \quad (1.3)$$

is given, then we select the Bayesian network consistent with this order; it is unique among the eight Markov-equivalent candidates. Nevertheless, having only the order (1.3) is insufficient to recover Fig. 1.2a.

Therefore, to learn a causal network from data, we need the following two stages:

Searching variable order Under certain assumptions, determine a variable order and thereby identify a single member among the Markov-equivalent Bayesian networks (Chap. 6).

Searching dependencies Determine the dependencies among variables and narrow down to the (multiple) Markov-equivalent Bayesian networks (Chaps. 5 and 8).

1.2 Searching Variable Order

Given data, standard statistical methods cannot identify the variable order. Since

$$P(X, Y) = P(X)P(Y \mid X) = P(X \mid Y)P(Y),$$

we cannot tell which factorization is appropriate. In this book, we assume the following for order discovery. Let X, Y be zero-mean continuous random variables. We posit that there exists a constant $a \neq 0$ and a random variable e independent of X such that

$$Y = aX + e. \tag{1.4}$$

When this holds, X is the cause and Y is the effect. By symmetry, swapping X and Y gives the following: There exists a constant $a' \neq 0$ and a random variable e' independent of Y such that

$$X = a'Y + e'. \tag{1.5}$$

When this holds, Y is the cause and X is the effect. This is the *additive noise model*. It is the criterion underlying linear non-Gaussian acyclic model (LiNGAM) discussed in this book, originally proposed by H. Kano in 2002.

You may wonder whether (1.4) and (1.5) could both hold simultaneously. The answer is clear:

Claim 1

(X, Y) is Gaussian $\iff$ (1.4) and (1.5) both hold.

We explain why in Chap. 6. Furthermore, assume at least one of (1.4) or (1.5) holds. Then Claim 1 implies:

Claim 2
If at least one of (1.4) or (1.5) holds, then

At least one of X, Y is non-Gaussian

$\iff$ The causal direction between X and Y is identifiable.

Thus, as long as we avoid the case where *both* X and Y are Gaussian, the cause–effect direction is identifiable; non-identifiability occurs only when both are Gaussian.

Based on this fact, proceed as follows: Given samples $(X, Y) = (x_1, y_1), \dots, (x_n, y_n)$, let $x^n := (x_1, \dots, x_n)$ and $y^n := (y_1, \dots, y_n)$. Define the regression coefficients a, a' via

$$c(x^n, y^n) := \frac{1}{n}\sum_{i=1}^{n} x_i y_i, \quad v(x^n) := \frac{1}{n}\sum_{i=1}^{n} x_i^2, \quad v(y^n) := \frac{1}{n}\sum_{i=1}^{n} y_i^2, \tag{1.6}$$

and set

$$a := \frac{c(x^n, y^n)}{v(x^n)}, \qquad a' := \frac{c(x^n, y^n)}{v(y^n)}.$$

Compute residual vectors (each of length n)

$$e := y^n - ax^n, \qquad e' := x^n - a'y^n.$$

Since we assume zero means, we do not subtract sample means $\overline{x} := n^{-1}\sum_i x_i$ and $\overline{y} := n^{-1}\sum_i y_i$ in (1.6).

We then test which pair is independent, (e, x^n) or (e', y^n)—that is, whether (1.4) or (1.5) holds. By Claim 2, we can determine which is the cause and which is the effect, provided the sample size n is sufficiently large.

This requires a test of independence. If $X \perp\!\!\!\perp Y$, then their correlation is zero, but not conversely. In our construction, both $c(e, x^n)$ and $c(e', y^n)$ equal zero; for example,

$$c(e, x^n) = \frac{1}{n}\sum_{i=1}^{n} e_i x_i = \frac{1}{n}\sum_{i=1}^{n}(y_i - ax_i)x_i = c(x^n, y^n) - a\, v(x^n) = 0.$$

Similarly $c(e', y^n) = 0$. When (X, Y) is Gaussian, independence is equivalent to zero correlation; this aligns with Claim 2.

Therefore, independence testing based on (near-)zero correlation is inadequate: we would fail to detect dependence whenever correlation happens to be zero. Such tests are meaningful only under joint Gaussianity and are inappropriate here.

A widely used alternative is Hilbert–Schmidt Independence Criterion *(HSIC)*. Let $k : \mathcal{X} \times \mathcal{X} \to \mathbb{R}$ be symmetric ($k(x, y) = k(y, x)$) and positive definite (for any $n \geq 1$ and $x_1, \dots, x_n \in \mathcal{X}$, the matrix $(k(x_i, x_j))$ is positive semidefinite). We call such k a (positive definite) kernel. To assess independence of random variables $X \in \mathcal{X}$ and $Y \in \mathcal{Y}$, we prepare kernels $k_{\mathcal{X}}$ and $k_{\mathcal{Y}}$ and examine the independence of the feature functions $k_{\mathcal{X}}(X, \cdot)$ and $k_{\mathcal{Y}}(Y, \cdot)$. HSIC captures nonlinear relations and has strong power in this setting. We study it in detail in Chap. 4.

This procedure is known as direct linear non-Gaussian acyclic model (LiNGAM). *LiNGAM* was proposed by Shimizu et al. in 2006 and has several variants. The original approach, ICA-LiNGAM, is based on independent component analysis. We introduce both, but focus on direct LiNGAM for its intuitive clarity.

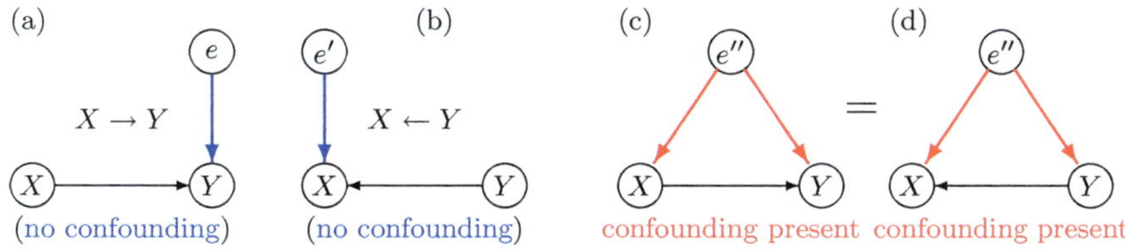

Fig. 1.4 Panel (**a**) represents, by a Bayesian network, the dependency when there exist constants a, a' such that $Y = aX + e$ with $X \perp\!\!\!\perp e$, and panel (**b**) represents the case $X = a'Y + e'$ with $Y \perp\!\!\!\perp e'$. In each case, there is a collider at Y or X, respectively. In this situation, noise is added only to either X or Y (i.e., there is no confounding). On the other hand, panels (**c**) and (**d**) show the case where noise is added to both X and Y (i.e., there is confounding). Either of the Markov equivalent Bayesian networks (**c**) or (**d**) can be used to represent this

The dependency patterns (1.4) and (1.5) correspond to Figs. 1.4a and b, where noise enters only one of X or Y. When neither (1.4) nor (1.5) holds, the situation is like Figs. 1.4c and d: noise affects both X and Y. We call this *confounding*. Standard LiNGAM assumes no confounding; we also discuss how to handle confounding in this book.

For multiple variables, consider X, Y, Z. There are six possible orders. If $X \to Y \to Z$, then

$$X = e_X, \qquad Y = aX + e_Y, \qquad Z = bX + cY + e_Z,$$

and we ask whether there exist constants a, b, c such that e_X, e_Y, e_Z are mutually independent.

Proceed from the upstream variable. Given samples $x^n = (x_1, \ldots, x_n)$, $y^n = (y_1, \ldots, y_n)$, $z^n = (z_1, \ldots, z_n)$, define

$$b := \frac{c(x^n, y^n)}{v(x^n)}, \qquad c := \frac{c(x^n, z^n)}{v(x^n)},$$

and compute residuals

$$y_x^n := y^n - bx^n, \qquad z_x^n := z^n - cx^n.$$

Test whether x^n is independent of $\{y_x^n, z_x^n\}$ (the latter can be viewed as an $n \times 2$ matrix). Do likewise for Y and Z: test y^n vs. $\{z_y^n, x_y^n\}$ and z^n vs. $\{x_z^n, y_z^n\}$. For convenience, you may compute HSIC for each of the three pairs and choose the one with the smallest value.

Suppose X is found to be the most upstream. Then set

$$b := \frac{c(y_x^n, z_x^n)}{v(z_x^n)}, \qquad c := \frac{c(y_x^n, z_x^n)}{v(y_x^n)},$$

and residuals

$$y^n_{xz} := y^n_x - bz^n_x, \qquad z^n_{xy} := z^n_x - cy^n_x.$$

Test independence between y^n_x and z^n_{xy}, and between z^n_x and y^n_{xz}. If, say, the former are independent (so Y is upstream of Z), then the overall order is $X \to Y \to Z$.

In this way, for any number of variables, we can determine the order from residual-based independence tests while removing the influence of already-selected upstream variables.

1.3 Searching Dependencies

Once the variable order is fixed, we only need to determine, for each variable, which upstream variables it depends on (its parent set).

In the Asia dataset, given the order (1.3), A has an empty parent set; T has either $\{A\}$ or $\emptyset$; S has one of $\{A, T\}$, $\{A\}$, $\{T\}$, or $\emptyset$; and so on. For each variable, we select a subset of the upstream variables.

We then score candidate parent sets using *information criteria* (Akaike Information Criterion—AIC or Bayesian Information Criterion—BIC) or *marginal likelihood*, and choose the best parent set.

A brief detour: very few people doubt "Newton's laws of motion." Newtonian mechanics has explained real-world motion with remarkable accuracy for centuries. One reason such laws are widely accepted is their *parsimony*: a few simple laws explain many phenomena.

Model selection in statistics follows a similar philosophy: in addition to goodness of fit, we value how simple the model is. Concretely, for a model M, we evaluate lack of fit by the negative log-likelihood H_M and complexity by the number of parameters d_M (smaller sums are better). Information criteria and marginal likelihood balance these two to choose a model with high explanatory power yet "just-right" complexity.

Akaike's Information Criterion (AIC) and Bayesian Information Criterion (BIC) weight H_M and d_M with ratios 1:1 and 1:$\frac{1}{2}\log n$, respectively, where n is the sample size:

$$AIC_M = H_M + d_M, \qquad BIC_M = H_M + \frac{d_M}{2}\log n. \tag{1.7}$$

We then select the model M minimizing each score. Suppose, for variable S, the parent set M induces m conditioning states (e.g., with parents A, T, there are four combinations). In Asia, the candidates for the parent set of S are $\{A, T\}$, $\{A\}$, $\{T\}$, and {}, with $m = 4, 2, 2, 1$, respectively.

If the parent set is empty, let θ be $P(S = 1)$ and c its count in the data. The likelihood is

$$\prod_{i=1}^{n} P(x_i \mid \theta) = \theta^c (1-\theta)^{n-c}.$$

Differentiating the log-likelihood w.r.t. θ and setting to zero yields the Maximum Likelihood Estimate (MLE)

$$\frac{\partial}{\partial \theta} \log\{\theta^c (1-\theta)^{n-c}\} = \frac{c}{\theta} - \frac{n-c}{1-\theta} = 0 \Rightarrow \theta = \frac{c}{n}.$$

The negative log-likelihood is

$$\mathcal{H}_2(p) := -p \log p - (1-p)\log(1-p),$$

so $H_M = n\mathcal{H}_2(c/n)$. For $m \geq 2$, let n_j be the frequency of state j and c_j, the frequency of $S = 1$ within that state; then $H_M = \sum_{j=1}^{m} n_j \mathcal{H}_2(c_j/n_j)$. From the dataset, we obtain the following counts:

A	T	$S=0$	$S=1$
0	0	2448	2468
0	1	21	21
1	0	15	25
1	1	1	1

A	$S=0$	$S=1$
0	2469	2489
1	16	26

T	$S=0$	$S=1$
0	2463	2493
1	22	22

$S=0$	$S=1$
2485	2515

Since each variable is binary and each state contributes one probability parameter, when the parent set of M has m states, $d_M = m$. The resulting AIC and BIC are:

M	H_M	d_M	AIC_M	BIC_M
$\{A, T\}$	4997.69	4	5001.69	5014.72
$\{A\}$	4997.96	2	4999.96	5006.48
$\{T\}$	4999.51	2	5001.51	5008.03
$\{\}$	4999.50	1	5000.50	5003.76

Thus, with AIC, we draw a directed edge $A \rightarrow S$, while with BIC, we draw no incoming edge to S. Although Asia is entirely binary, H_M and d_M are defined similarly for general multinomial variables (see Chap. 8).

Given the order, performing this parent selection step for every variable yields a Bayesian network.

For continuous variables, proceed similarly. Let $(X_j)_{j\in\pi}$ be the parent set of X_k. From the samples $x_{i,j}$ $(j \in \pi)$ and $x_{i,k}$ $(i = 1, \dots, n)$,

$$\hat{\sigma}^2 = \min_{\beta_j,\, j\in\pi} \left\{ \sum_{i=1}^{n} \left(x_{i,k} - \sum_{j\in\pi} x_{i,j} \right)^2 \right\},$$

and set $H_M = 0.5 \log \hat{\sigma}^2$ and $d_M = |\pi|$.

Some references define AIC as minimizing $2H_M + 2d_M$ or equivalently maximizing $-2H_M - 2d_M$; the optimizer M is the same. We use the notation in (1.7) throughout.

We also consider dependency search via marginal likelihood. For i.i.d., length-3 binary sequences (eight patterns), assign

$$P(x_1x_2x_3 \mid \theta) = \theta^k (1-\theta)^{3-k}, \quad 0 \le k \le 3,$$

and place the prior

$$\frac{1}{\pi\sqrt{\theta(1-\theta)}}, \qquad 0 \le \theta \le 1.$$

Then

$$\int_0^1 \theta^k (1-\theta)^{3-k} \cdot \frac{1}{\pi\sqrt{\theta(1-\theta)}}\, d\theta = \int_0^1 \theta^{k-0.5}(1-\theta)^{2.5-k}\, d\theta$$

evaluates to $3/8$ for $k = 0, 3$ and $1/8$ for $k = 1, 2$. This is the marginal likelihood. With a chosen prior for parameters, we can compute it for any length n sequence x^n, and likewise for paired data (x_i, y_i). Denote $Q(x^n, y^n)$ as the marginal likelihood for (X, Y); the definition extends to discrete and continuous variables.

Using marginal likelihoods, with prior probability $0 < r < 1$ for independence, one can show that, with probability 1 as $n \to \infty$,

$$r\, Q(x^n) Q(y^n) \le (1-r)\, Q(x^n, y^n)$$

if and only if X and Y are independent. The asymptotic performance does not depend on r as long as $r \notin \{0, 1\}$.

Moreover, when selecting the parent set of Z from subsets of $\{X, Y\}$, compare

$$\frac{Q(x^n, y^n, z^n)}{Q(x^n, y^n)}, \qquad \frac{Q(x^n, z^n)}{Q(x^n)}, \qquad \frac{Q(y^n, z^n)}{Q(y^n)}, \qquad Q(z^n),$$

and choose among $\{X, Y\}$, $\{X\}$, $\{Y\}$, or $\emptyset$ according to which is largest. Crucially, one can prove that such parent selection converges to the true dependency as

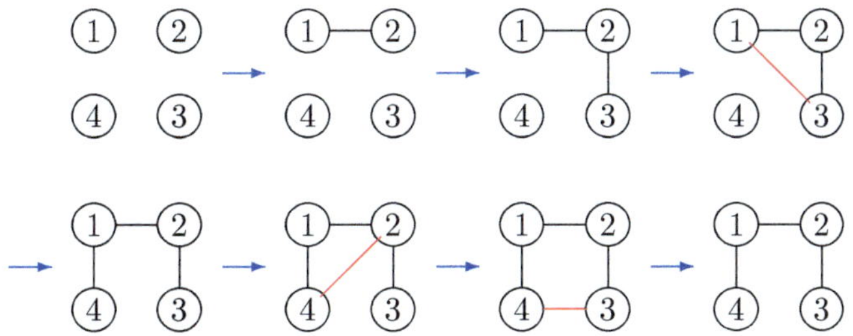

Fig. 1.5 For the random variables X_1, X_2, X_3, X_4, suppose the mutual information satisfy $I(X_1, X_2) > I(X_2, X_3) > I(X_1, X_3) > I(X_1, X_4) > I(X_2, X_4) > I(X_3, X_4)$. Edges $\{1, 2\}$, $\{2, 3\}$, $\{1, 3\}$, $\{1, 4\}$, $\{2, 4\}$, and $\{3, 4\}$ are added in this order. When attempting to add $\{1, 3\}$, a cycle is created, so $\{1, 3\}$ is not included. The same applies to $\{2, 4\}$ and $\{3, 4\}$

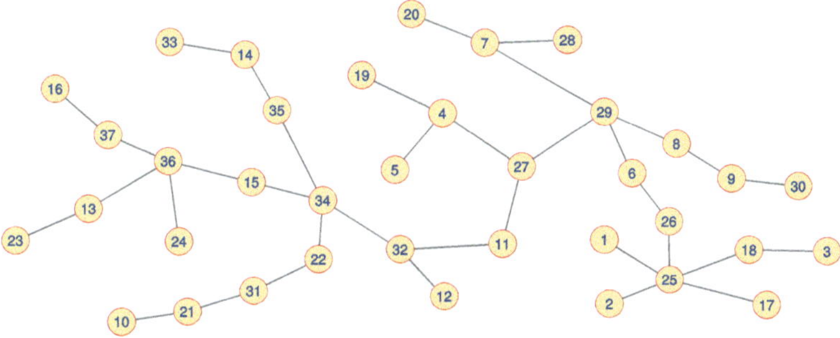

Fig. 1.6 Alarm network ($p = 37$). Computing the exact optimum by information criteria or marginal likelihood takes over 24 hours already at $p = 30$, about two days at $p = 31$, and four days at $p = 32$—exponential in p

$n \to \infty$. BIC approximates the negative log of the marginal likelihood and thus has similar asymptotic behavior (Fig. 1.5).

A drawback of estimating the exact BN structure by information criteria or marginal likelihood is the enormous computation time–exponential in p (Fig. 1.6). One can restrict the search space (e.g., with hill-climbing) while scoring by an information criterion.

Another approach is the *Chow–Liu algorithm*, which constructs a tree from the pairwise mutual information $I(X_i, X_j)$ for $1 \le i < j \le p$. Add undirected edges in order of decreasing $I(X_i, X_j)$, skipping any edge that would create a cycle (Fig. 1.5). The resulting tree maximizes the sum of edge mutual information and, moreover, is guaranteed to minimize the Kullback–Leibler divergence from the true distribution among all trees (Chap. 8).

From marginal likelihoods, we can also define an estimator of mutual information

$$I(X, Y) = \sum_k \sum_l P(X = k, Y = l) \log \frac{P(X = k, Y = l)}{P(X = k)P(Y = l)}.$$

Given counts $n_{k,l}$ for $(X, Y) = (k, l)$ (with $n_{k,\cdot} = \sum_l n_{k,l}$ and $n_{\cdot,l} = \sum_k n_{k,l}$), the MLE-based estimator

$$I_n := \sum_k \sum_l \frac{n_{k,l}}{n} \log \frac{(n_{k,l}/n)}{(n_{k,\cdot}/n)\,(n_{\cdot,l}/n)}$$

is commonly used; indeed $I_n \to I(X, Y)$ (consistency). Alternatively, using marginal likelihoods, define

$$J_n := \frac{1}{n} \log \frac{Q(x^n, y^n)}{Q(x^n)Q(y^n)} \approx I_n - \frac{(\alpha - 1)(\beta - 1)}{2n} \log n,$$

where α, β are the numbers of categories of X and Y. This yields not only consistency but also *detection* of independence: with probability 1,

$$X \perp\!\!\!\perp Y \Longleftrightarrow J_n \leq 0.$$

Thus, if $J_n \leq 0$, we do not connect the pair-generalizing Chow–Liu to produce a forest rather than a tree. When $X \perp\!\!\!\perp Y$, $I_n \to 0$, but remains positive, so I_n alone cannot decide independence. For continuous variables, an analogous method applies; set

$$J_n = I_n - \frac{1}{2n} \log n.$$

This forest construction algorithm scales roughly with p^2 rather than exponentially in p.

1.4 Graphical Models and Conditional Independence

At the outset, I mentioned that arrows in a Bayesian network do not necessarily represent causal directions. Let me clarify this point to close the chapter.

Bayesian networks admit an equivalent definition: they are DAGs that jointly encode the conditional independences among $X_1, \ldots, X_p$. Markov networks likewise encode conditional independences with an undirected graph.

Whereas independence may be intuitive, conditional independence can be harder to grasp. Events A and B are independent if $P(A, B) = P(A)P(B)$. Events A and C are conditionally independent given B if

$$P(A, B, C) = \frac{P(A, B)P(B, C)}{P(B)}. \tag{1.8}$$

As an example, let A, B, C denote rain on January 1, 2, and 3. Since weather evolves continuously, the weather on the 3rd may be predictable from that on the 2nd without knowing the 1st. We write this as $P(C \mid B) = P(C \mid A, B)$ or (1.8). Conditional independence is defined similarly for (sets of) random variables, and we can represent it graphically.

First consider an undirected graph G. For disjoint index sets $A, B, C \subset \{1, \ldots, p\}$, if every path from any $i \in A$ to any $j \in C$ passes through some vertex in B, then we say A and C are separated by B and write $A \perp\!\!\!\perp_G C \mid B$. Let $X_A := \{X_i\}_{i\in A}, X_B := \{X_j\}_{j\in B}, X_C := \{X_k\}_{k\in C}$. We aim to have

$$A \perp\!\!\!\perp_G C \mid B \Longrightarrow X_A \perp\!\!\!\perp X_C \mid X_B \tag{1.9}$$

hold as much as possible for the distribution of $X_1, \ldots, X_p$. If all pairs are connected, there is no separation, so (1.9) holds vacuously. As we remove edges, (1.9) may fail. The Markov network is the sparsest undirected graph G for which (1.9) still holds.

For Bayesian networks, a similar condition holds, but separation is defined somewhat differently for DAGs. In Fig. 1.7a,b,c, separation is as in undirected graphs; in the collider case (d), however, separation flips depending on whether node ② or its descendants are conditioned on. The precise definition is given in Chap. 8; this notion is called *d-separation*. Using it, we obtain the conditional independences encoded by a BN. For large DAGs, reading $A \perp\!\!\!\perp_G C \mid B$ off the graph may be nontrivial (not mathematically difficult, but it requires practice).

Note that a single undirected graph or DAG typically encodes multiple conditional independences. Also, the sets of independences representable by Markov vs. Bayesian networks differ; an implication like (1.9) with "$\Longleftrightarrow$" may hold for a BN, but not for a Markov network, and vice versa.

So far, we have explored dependency discovery by optimizing a score such as an information criterion or marginal likelihood (the *score-based* approach). An alternative is the *constraint-based* approach, which constructs a BN by testing conditional independences from data.

A well-known constraint-based method is the *Peter–Clark (PC) algorithm*. It assumes (for the underlying distribution) that the "$\Longleftrightarrow$" in (1.9) holds (the *faithfulness* assumption). In brief (details in Chap. 5), we perform tests of the form

$$X_i \perp\!\!\!\perp X_j \mid (X_k)_{k\in\pi}, \qquad \pi \subseteq \{1, \ldots, p\} \setminus \{i, j\}, \tag{1.10}$$

Fig. 1.7 Representation of separations using DAGs. (**a**), (**b**), and (**c**) represent $1 \perp\!\!\!\perp 3 \mid 2$, while (**d**) represents $1 \perp\!\!\!\perp 3$

increasing $|\pi|$ up to the largest set for which (1.10) still holds. Under faithfulness, all conditional independences are of this form (Chap. 5). After these tests, for each $i \neq j$ we obtain a maximal separating set $\pi(i, j)$ with $X_i \perp\!\!\!\perp X_j \mid \pi(i, j)$. However, this yields only a Markov-equivalence class of BNs; without variable-order discovery, causal discovery remains incomplete.

As with independence testing, conditional independence cannot be reliably inferred from small correlations alone. We therefore apply *KCI* (Kernel-based conditional independence test), an extension of HSIC (Chap. 4).

In this chapter, we presented an intuitive overview of causal discovery. From the next chapter, we return to the basics of probability and statistics and formally develop the prerequisites for causal discovery.

The causal discovery treated in this book deals with graphical models and differs from the causal inference in the sense of Donald Rubin. It is often referred to as Pearl-style causal inference [20].

Chapter 2
Foundations of Probability and Statistics

In this chapter, as the groundwork for learning causal discovery with graphical models, we introduce the basic concepts of probability and statistics. We begin by defining fundamental terms such as events, probabilities, and random variables and then introduce representative probability distributions: the binomial, normal, Poisson, and Gamma distributions. We also discuss concepts used to describe relations among multiple random variables—joint distributions, independence, correlation coefficients, and covariance matrices.

We further present properties of the normal distribution—indispensable in multivariate analysis and causal inference—and its behavior under linear transformations, thereby laying the mathematical foundation for describing probabilistic structure. We briefly touch on the Wishart distribution and its inverse, showing their connection to Bayesian statistics.

In the second half, we introduce maximum likelihood estimation as a method to estimate model parameters from observed data and explain the framework of statistical testing. We define type I/II errors, significance level, and power and provide an overall understanding of hypothesis testing.

Below, we denote by $\mathbb{R}$, $\mathbb{Q}$, $\mathbb{Z}$, and $\mathbb{N}$ the sets of all real numbers, rational numbers, integers, and nonnegative integers, respectively. For positive integers l, m, n, we write $\mathbb{R}^{l\times m}$ and $\mathbb{R}^n$ for the set of real $l \times m$ matrices and the set of (column) vectors of length n. For $a < b$, the intervals (a, b), $(a, b]$, $[a, b)$, $[a, b]$ denote $\{x \in \mathbb{R} : a < x < b\}$, $\{x \in \mathbb{R} : a < x \leq b\}$, $\{x \in \mathbb{R} : a \leq x < b\}$, and $\{x \in \mathbb{R} : a \leq x \leq b\}$, respectively.

2.1 Random Variables

When rolling a die, one of the faces $1, 2, 3, 4, 5, 6$ occurs at random. A specific condition regarding the outcome (e.g., "an even face occurs" or "a 1 or 2 occurs") is called an *event*. We call the entire set, such as $\{1, 2, 3, 4, 5, 6\}$, the *sample space* and

J. Suzuki, *Graphical Models and Causal Discovery with R*,
https://doi.org/10.1007/978-981-95-4267-3_2

denote it by Ω. We represent events as subsets like $\{2, 4, 6\}$ or $\{1, 2\}$ (there are 2^6 such subsets). The family $\mathcal{F}$ whose elements are these events is called the σ-algebra (collection of events). Thus,

$$\Omega = \{1, 2, 3, 4, 5, 6\}, \qquad \mathcal{F} = \underbrace{\{\{\}, \{1\}, \{2\}, \{3\}, \{4\}, \{5\}, \{6\}, \ldots, \{1, 2, 3, 4, 5, 6\}\}}_{2^6 \text{ elements}},$$

where $\{\}$ denotes the empty set. Each event $A \in \mathcal{F}$ is a subset of Ω, and $\mathcal{F}$ must be closed under intersection ($\cap$), union ($\cup$), and complementation ($\bar{\ }$), namely,

$$A \in \mathcal{F} \Longrightarrow \bar{A} \in \mathcal{F}, \qquad A, B \in \mathcal{F} \Longrightarrow \begin{cases} A \cap B \in \mathcal{F} \\ A \cup B \in \mathcal{F} \end{cases}$$

A map $P : \mathcal{F} \to [0, 1]$ is called a *probability* if it assigns to each event $A \in \mathcal{F}$ a value $P(A)$ satisfying

1. $P(A) \geq 0$ for all $A \in \mathcal{F}$
2. $P(\Omega) = 1$
3. For pairwise disjoint $A_1, A_2, \ldots$

$$P\left(\bigcup_{i=1}^{\infty} A_i\right) = \sum_{i=1}^{\infty} P(A_i).$$

If the die is biased and we estimate probabilities empirically, we must assign 2^6 probabilities satisfying the three axioms above.

From these axioms, it follows that

$$P(\{\}) = 0 \tag{2.1}$$

and, for all $A, B \in \mathcal{F}$,

$$P(A \cup B) = P(A) + P(B) - P(A \cap B) \tag{2.2}$$

(Problem 1).

In what follows, we assume the triple $(\Omega, \mathcal{F}, P)$ is given.[1]

Return to the die. Define a variable X that takes $X = 0$ for even outcomes and $X = 1$ for odd outcomes:

$$\{1, 2, 3, 4, 5, 6\} \ni \omega \mapsto X(\omega) = \begin{cases} 0, & \omega = 2, 4, 6 \\ 1, & \omega = 1, 3, 5 \end{cases}$$

[1] Called a **probability space**.

A map $X : \Omega \to \mathbb{R}$ of this kind is called a *random variable* and is used to define events; e.g., $X = 1$ corresponds to the event $\{1, 3, 5\}$. We write random variables in uppercase and their realizations in lowercase.

We distinguish *finite* vs. *infinite* sets, and among infinite sets, *countable* vs. *uncountable*. A countable set can be put in one-to-one correspondence with $\mathbb{N}$. $\mathbb{Q}$ is countable, and $\mathbb{R}$ is uncountable. Finite or countable sets are said to have *at most countably many* elements.

Without loss of generality, we assume random variables take real values. A random variable taking values in a finite set (identified with nonnegative integers up to some bound) or in a countable set (identified with nonnegative integers) is called a *discrete random variable*. In the die example, X is discrete with values $\{0, 1\}$. In Examples 1 and 2, X takes finitely and infinitely many values, respectively. We may assume X takes values $0, 1, 2, \ldots$, and if it takes only m values, set $p_m = p_{m+1} = \cdots = 0$. From the axioms, $p_k \geq 0$ and

$$\sum_{k=0}^{\infty} p_k = 1 \tag{2.3}$$

are required.

Example 1 (Binomial Distribution) Suppose a wrestler wins a bout with probability $0 < p < 1$ independently of the opponent. After $n(\geq 1)$ bouts, the probability of k wins and $n - k$ losses $(0 \leq k \leq n)$ is

$$p_k = \frac{n!}{k!(n-k)!} p^k (1-p)^{n-k}.$$

Indeed, by the binomial theorem,

$$1 = (p + 1 - p)^n = \sum_{k=0}^{n} \frac{n!}{k!(n-k)!} p^k (1-p)^{n-k} = \sum_{k=0}^{n} p_k,$$

so (2.3) holds. ■

Example 2 (Poisson Distribution) Let $X = 0, 1, 2, \ldots$ denote the number of customers waiting at a store. Often we assume for some $\lambda > 0$

$$p_k = e^{-\lambda} \frac{\lambda^k}{k!}.$$

Using the Maclaurin expansion $e^x = \sum_{k=0}^{\infty} x^k / k!$ with $x = \lambda$ yields (2.3). ■

When, for all $x \in \mathbb{R}$, the cumulative probability $P(X \leq x)$ can be written with an integrable function f_X over its domain as

$$P(X \leq x) = \int_{-\infty}^{x} f_X(t)\, dt, \tag{2.4}$$

f_X is called the *probability density function* (pdf) of X, and X is a *continuous random variable*. From the axioms, $f_X(x) \geq 0$ and

$$1 = \lim_{x \to \infty} P(X \leq x) = \int_{-\infty}^{\infty} f_X(x)\, dx \tag{2.5}$$

must hold.

Change of variables is often needed. In two dimensions, for $g, h : \mathbb{R}^2 \to \mathbb{R}$,

$$\int_{x_1}^{x_2} \int_{y_1}^{y_2} f(x, y)\, dx\, dy = \int_{u_1}^{u_2} \int_{v_1}^{v_2} f\big(g(u, v), h(u, v)\big)\, |J|\, du\, dv,$$

where

$$x_1 = g(u_1, v_1),\ x_2 = g(u_2, v_2),\ y_1 = h(u_1, v_1),\ y_2 = h(u_2, v_2),$$

and J is the *Jacobian*

$$J = \det \begin{bmatrix} \dfrac{\partial g(u, v)}{\partial u} & \dfrac{\partial g(u, v)}{\partial v} \\ \dfrac{\partial h(u, v)}{\partial u} & \dfrac{\partial h(u, v)}{\partial v} \end{bmatrix} = \frac{\partial g}{\partial u} \cdot \frac{\partial h}{\partial v} - \frac{\partial g}{\partial v} \cdot \frac{\partial h}{\partial u}.$$

Example 3 (Standard Normal Distribution) Let

$$f_X(x) = \frac{1}{\sqrt{2\pi}} \exp\left(-\frac{x^2}{2}\right)$$

be the pdf of continuous X; we say X follows the *standard normal distribution*. For $r > 0, 0 \leq \theta \leq \pi/2$, set $x = r \cos\theta$, $y = r \sin\theta$. The Jacobian is

$$J = \det \begin{bmatrix} \dfrac{\partial x}{\partial r} & \dfrac{\partial x}{\partial \theta} \\ \dfrac{\partial y}{\partial r} & \dfrac{\partial y}{\partial \theta} \end{bmatrix} = \det \begin{bmatrix} \cos\theta & -r\sin\theta \\ \sin\theta & r\cos\theta \end{bmatrix} = r > 0.$$

Since $x^2 + y^2 = r^2$,

$$\left\{ \int_{-\infty}^{\infty} \exp\left(-\frac{x^2}{2}\right) dx \right\}^2 = \int_{-\infty}^{\infty} \int_{-\infty}^{\infty} \exp\left(-\frac{x^2 + y^2}{2}\right) dx\, dy$$

$$= 4\int_0^\infty \int_0^{\pi/2} \exp\left(-\frac{r^2}{2}\right) r\, dr\, d\theta = 2\pi \tag{2.6}$$

(Problem 2), which gives (2.5). ■

Next, for $\alpha > 0$, define the *Gamma function*

$$\Gamma(\alpha) := \int_0^\infty t^{\alpha-1} e^{-t}\, dt. \tag{2.7}$$

For $\alpha > 0$,

$$\Gamma(\alpha + 1) = \alpha\, \Gamma(\alpha). \tag{2.8}$$

In particular, for nonnegative integers n, $\Gamma(n+1) = n!$ (Problem 3); also $\Gamma(1/2) = \sqrt{\pi}$ (Problem 3). If

$$f_X(x) = \frac{1}{\Gamma(a)} b^a x^{a-1} e^{-bx}, \quad x > 0, \tag{2.9}$$

then X has a *Gamma distribution*, written $X \sim G(a, b)$. Setting $a = n/2$ and $b = 1/2$ yields the *chi-square distribution* with n degrees of freedom, $X \sim \chi_n^2$, having density

$$f_X(x) = \frac{1}{2^{n/2}\Gamma(n/2)} x^{n/2-1} e^{-x/2}, \quad x > 0. \tag{2.10}$$

For $a, b > 0$, define the *Beta function*

$$B(a, b) := \int_0^1 \theta^{a-1}(1-\theta)^{b-1}\, d\theta.$$

Without proof, it relates to Γ as

$$B(a, b) = \frac{\Gamma(a)\Gamma(b)}{\Gamma(a+b)}. \tag{2.11}$$

The *Beta distribution* with density

$$f_X(x) = \frac{x^{a-1}(1-x)^{b-1}}{B(a, b)}, \quad 0 < x < 1$$

is written $X \sim Be(a, b)$.

Not all random variables are discrete or continuous.

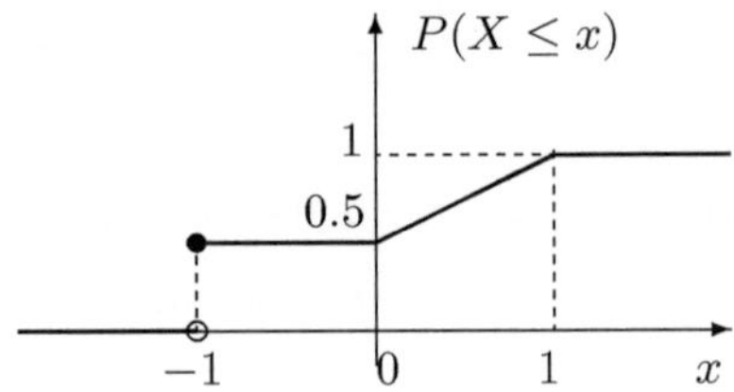

Fig. 2.1 For the random variable X in Example 5, there does not exist a probability density function f_X: $P(X \le x)$ is discontinuous at $x = -1$

Example 4 Let $X = -1$ with probability $1/2$, and $0 \le X \le 1$ with probability $1/2$ having density $f_X(x) = 1/2$ on $[0, 1]$. Then

$$P(X \le x) = \begin{cases} 0, & x < -1 \\ \frac{1}{2}, & -1 \le x < 0 \\ \frac{1+x}{2}, & 0 \le x < 1 \\ 1, & 1 \le x \end{cases}$$

X takes uncountably many values, but no f_X exists satisfying (2.4) (Fig. 2.1); thus, X is not continuous. ∎

Random variables that are neither purely discrete nor continuous are inconvenient; hence, when multiple variables appear, we handle either all-discrete or all-continuous cases.

A continuous function of a random variable is also a random variable.[2] For example, scalar multiples, shifts, and squares of X are random variables.

Define the *expectation* $\mathbb{E}[X]$ of X[3] by

$$\mathbb{E}[X] := \begin{cases} \sum_{k=0}^{\infty} k\, P(X = k), & X \text{ discrete} \\ \int_{-\infty}^{\infty} x f_X(x)\, dx, & X \text{ continuous.} \end{cases}$$

Let $\mu = \mathbb{E}[X]$. Then $(X - \mu)^2$ is a random variable, and its expectation $\mathbb{V}[X] = \mathbb{E}[(X - \mu)^2]$ is the *variance* of X:

$$\mathbb{V}[X] := \begin{cases} \sum_{k=0}^{\infty} (k - \mu)^2 P(X = k), & X \text{ discrete} \\ \int_{-\infty}^{\infty} (x - \mu)^2 f_X(x)\, dx, & X \text{ continuous.} \end{cases}$$

[2] In general, measurability is required so that the induced events belong to the predetermined σ-algebra $\mathcal{F}$ (measurability).

[3] For continuous X, defined when $\int_{-\infty}^{\infty} |x| f_X(x)\, dx < \infty$.

For $a, b \in \mathbb{R}$,

$$\mathbb{E}[aX + b] = a\,\mathbb{E}[X] + b \tag{2.12}$$

$$\mathbb{V}[aX + b] = a^2\,\mathbb{V}[X] \tag{2.13}$$

$$\mathbb{V}[X] = \mathbb{E}[X^2] - \{\mathbb{E}[X]\}^2 \tag{2.14}$$

(Problem 4).

Example 5 (Normal Distribution) For standard normal X, $f_X(x)$ is even and $xf_X(x)$ is odd, so $\mathbb{E}[X] = 0$. Using (2.5),

$$\int_{-\infty}^{\infty} x^2 f_X(x)\,dx = \int_{-\infty}^{\infty} x \cdot \{-f_X(x)\}'\,dx = [x \cdot \{-f_X(x)\}]_{-\infty}^{\infty} + \int_{-\infty}^{\infty} f_X(x)\,dx = 1,$$

so $\mathbb{V}[X] = 1$. For $\sigma > 0$, $\mu \in \mathbb{R}$, with $y = \sigma x + \mu$ and $t = \sigma s + \mu$,

$$\int_{-\infty}^{x} \frac{1}{\sqrt{2\pi}} e^{-s^2/2}\,ds = \int_{-\infty}^{y} \frac{1}{\sqrt{2\pi}} \exp\left\{-\frac{(t-\mu)^2}{2\sigma^2}\right\} \frac{dt}{\sigma},$$

so the pdf of $Y = \sigma X + \mu$ is

$$f_Y(y) = \frac{1}{\sqrt{2\pi\sigma^2}} \exp\left\{-\frac{(y-\mu)^2}{2\sigma^2}\right\}. \tag{2.15}$$

By (2.12)–(2.13), $\mathbb{E}[Y] = \mu$ and $\mathbb{V}[Y] = \sigma^2$. This is the *normal distribution*, written as $Y \sim N(\mu, \sigma^2)$. ■

2.2 Multivariate Distributions

For events A, B with $P(B) > 0$,

$$P(A \mid B) := \frac{P(A \cap B)}{P(B)}$$

is the *conditional probability* of A given B. If

$$P(A \cap B) = P(A)P(B),$$

A and B are *independent*.

Similarly, for any $x, y \in \mathbb{R}$, we say random variables X, Y are independent if

$$P(X \le x, Y \le y) = P(X \le x)P(Y \le y). \tag{2.16}$$

In the discrete case (Problem 5), for all $a, b \in \mathbb{N}$,

$$P(X = a, Y = b) = P(X = a)\, P(Y = b). \tag{2.17}$$

In the continuous case, this is equivalent (w.p.1) to

$$f_{XY}(x, y) = f_X(x)\, f_Y(y), \tag{2.18}$$

where f_{XY} is the joint pdf satisfying

$$P(X \le x, Y \le y) = \int_{-\infty}^{x}\int_{-\infty}^{y} f_{XY}(s, t)\, ds\, dt.$$

For a function $h(X, Y)$,

$$\mathbb{E}_{XY}[h(X, Y)] := \begin{cases} \displaystyle\sum_{k=0}^{\infty}\sum_{l=0}^{\infty} h(k, l) P(X = k, Y = l), & \text{discrete} \\ \displaystyle\int_{-\infty}^{\infty}\int_{-\infty}^{\infty} h(x, y) f_{XY}(x, y)\, dx\, dy, & \text{continuous} \end{cases}$$

denotes the expectation.[4] With means μ_X, μ_Y,

$$\text{cov}(X, Y) := \mathbb{E}_{XY}[(X - \mu_X)(Y - \mu_Y)]$$

is the *covariance*, and when variances are nonzero,

$$\rho(X, Y) := \frac{\text{cov}(X, Y)}{\sqrt{\mathbb{V}[X]\mathbb{V}[Y]}}$$

is the *correlation coefficient*, always in $[-1, 1]$ (Problem 6). The *covariance matrix* is

$$\Sigma = \begin{bmatrix} \mathbb{V}[X] & \text{cov}(X, Y) \\ \text{cov}(X, Y) & \mathbb{V}[Y] \end{bmatrix} = \begin{bmatrix} \mathbb{V}[X] & \rho(X, Y)\sqrt{\mathbb{V}[X]\mathbb{V}[Y]} \\ \rho(X, Y)\sqrt{\mathbb{V}[X]\mathbb{V}[Y]} & \mathbb{V}[Y] \end{bmatrix}.$$

For a linear transform with $A \in \mathbb{R}^{2\times 2}$, $B \in \mathbb{R}^2$,

[4] We may write $\mathbb{E}_X[\cdot]$ or $\mathbb{E}_Y[\cdot]$ to emphasize which variable the expectation is taken over.

$$A\begin{bmatrix} X \\ Y \end{bmatrix} + B,$$

the mean and covariance are

$$A\begin{bmatrix} \mathbb{E}[X] \\ \mathbb{E}[Y] \end{bmatrix} + B, \qquad A\Sigma A^{\top} \tag{2.19}$$

(Problem 4).

Example 6 (Bivariate Normal) Let U, V be independent standard normals, so

$$f_{UV}(u, v) = f_U(u) f_V(v) = \frac{1}{2\pi} \exp\left(-\frac{u^2 + v^2}{2}\right).$$

Set

$$\sigma_X := \sqrt{\mathbb{V}[X]},\ \sigma_Y := \sqrt{\mathbb{V}[Y]},\ \mu_X := \mathbb{E}[X],\ \mu_Y := \mathbb{E}[Y],\ \rho_{XY} = \rho(X, Y),$$

and define

$$\begin{bmatrix} x \\ y \end{bmatrix} = A\begin{bmatrix} u \\ v \end{bmatrix} + \begin{bmatrix} \mu_X \\ \mu_Y \end{bmatrix}, \quad A = \begin{bmatrix} \sigma_X & 0 \\ \sigma_Y \rho_{XY} & \sigma_Y \sqrt{1 - \rho_{XY}^2} \end{bmatrix}.$$

Then

$$[x - \mu_X, y - \mu_Y](AA^{\top})^{-1} \begin{bmatrix} x - \mu_X \\ y - \mu_Y \end{bmatrix} = u^2 + v^2,$$

and

$$AA^{\top} = \begin{bmatrix} \sigma_X^2 & \sigma_X \sigma_Y \rho_{XY} \\ \sigma_X \sigma_Y \rho_{XY} & \sigma_Y^2 \end{bmatrix} =: \Sigma_{XY}.$$

As in Example 3, the Jacobian is

$$J = \det(A^{-1}) = \frac{1}{\sigma_X \sigma_Y \sqrt{1 - \rho_{XY}^2}} = \frac{1}{\sqrt{\det \Sigma_{XY}}}.$$

Thus,

$$f_{XY}(x, y) = \frac{1}{2\pi \sqrt{\det \Sigma_{XY}}} \exp\left\{-\frac{1}{2}[x - \mu_X, y - \mu_Y]\Sigma_{XY}^{-1} \begin{bmatrix} x - \mu_X \\ y - \mu_Y \end{bmatrix}\right\},$$

with mean (μ_X, μ_Y) and covariance Σ_{XY}. ∎

In general, if independent $U_1, \ldots, U_p$ are standard normal and

$$\begin{bmatrix} X_1 \\ \vdots \\ X_p \end{bmatrix} = A \begin{bmatrix} U_1 \\ \vdots \\ U_p \end{bmatrix} + \mu$$

for some $A \in \mathbb{R}^{p \times p}$ and $\mu \in \mathbb{R}^p$, then $[X_1, \ldots, X_p]^\top \sim N(\mu, \Sigma)$ with $\Sigma = AA^\top$.

A symmetric matrix $A \in \mathbb{R}^{p \times p}$ is *nonnegative definite* if $x^\top Ax \geq 0$ for all x, and *positive definite* if $x^\top Ax > 0$ for all $x \neq 0$. A symmetric matrix is positive definite iff all eigenvalues are positive.

For any positive definite symmetric A, there exists a unique lower triangular L with positive diagonal entries such that $A = LL^\top$; this is the *Cholesky decomposition.*

Example 7 Covariance matrices are nonnegative definite. If $X_1, \ldots, X_p$ have mean $\mu_1, \ldots, \mu_p$ and covariance Σ, then for any $z \in \mathbb{R}^p$,

$$z^\top \Sigma z = \mathbb{E}\left[\left\{\sum_{i=1}^{p} z_i (X_i - \mu_i)\right\}^2\right] \geq 0.$$

Thus, there is a unique lower triangular A with positive diagonal such that $\Sigma = AA^\top$. ■

For continuous X, Y, the joint density f_{XY} contains all information. Marginals are

$$f_X(x) = \int_{-\infty}^{\infty} f_{XY}(x, y)\, dy, \qquad f_Y(y) = \int_{-\infty}^{\infty} f_{XY}(x, y)\, dx,$$

and the conditional density is

$$f_{Y|X}(y \mid x) := \frac{f_{XY}(x, y)}{f_X(x)}, \qquad \int_{-\infty}^{\infty} f_{Y|X}(y \mid x)\, dy = 1,$$

so

$$f_{XY}(x, y) = f_X(x)\, f_{Y|X}(y \mid x).$$

Example 8 (Marginalization of Bivariate Normals) If $X \sim N(\mu_X, \sigma_X^2)$, $Y \sim N(\mu_Y, \sigma_Y^2)$ with correlation ρ, then

$$Y \mid (X = x) \sim N\left(\mu_Y + \rho\frac{\sigma_Y}{\sigma_X}(x - \mu_X),\ (1 - \rho^2)\sigma_Y^2\right) \tag{2.20}$$

(Problem 7). ■

If X, Y are independent, then $\rho(X, Y) = 0$ and $\text{cov}(X, Y) = 0$. For continuous variables,

$$\iint (x - \mu_X)(y - \mu_Y) f_X(x) f_Y(y)\, dx\, dy$$
$$= \Big(\int (x - \mu_X) f_X(x)\, dx\Big)\Big(\int (y - \mu_Y) f_Y(y)\, dy\Big) = 0.$$

The discrete case is analogous. However, zero covariance does not imply independence.

Example 9 Let X be standard normal and $Y = X^2$. Then X, Y are not independent, but $\text{cov}(X, Y) = 0$ because $\mathbb{E}[X] = \mathbb{E}[X^3] = 0$ and $\mathbb{E}[Y] = \mathbb{E}[X^2] = 1$:

$$\text{cov}(X, Y) = \text{cov}(X, X^2) = \mathbb{E}[X(X^2 - 1)] = \mathbb{E}[X^3] - \mathbb{E}[X] = 0.$$

■

For the normal family, the converse holds.

Proposition 1 *For a bivariate normal distribution, zero covariance is equivalent to independence.*

Proof If $\text{cov}(X, Y) = 0$, then

$$\Sigma = \text{diag}(\sigma_X^2, \sigma_Y^2), \quad \Sigma^{-1} = \text{diag}\left(\sigma_X^{-2}, \sigma_Y^{-2}\right),$$

and

$$f_{XY}(x, y) = \frac{1}{\sqrt{2\pi\sigma_X^2}} \exp\left(-\frac{(x - \mu_X)^2}{2\sigma_X^2}\right) \cdot \frac{1}{\sqrt{2\pi\sigma_Y^2}} \exp\left(-\frac{(y - \mu_Y)^2}{2\sigma_Y^2}\right),$$

which factors into $f_X(x) f_Y(y)$; hence, X and Y are independent.

■

Finally, define the *characteristic function* of X by

$$\Psi_X(t) := \mathbb{E}[e^{iXt}].$$

If a pdf f_X exists,

$$\Psi_X(t) = \int_{-\infty}^{\infty} e^{ixt} f_X(x)\, dx,$$

a complex-valued function. Then

$$\Psi_X(0) = 1 \tag{2.21}$$

$$|\Psi_X(t)| \leq \mathbb{E}[|e^{iXt}|] = \mathbb{E}[1] = 1 \tag{2.22}$$

$$\Psi_X{}'(0) = i\,\mathbb{E}[X] \tag{2.23}$$

$$\Psi_X{}''(0) = -\,\mathbb{E}[X^2] \tag{2.24}$$

hold. For $X_1, \ldots, X_n$,

$$\Psi_{X_1,\ldots,X_n}(t_1, \ldots, t_n) := \mathbb{E}[e^{i(X_1t_1+\cdots+X_nt_n)}].$$

Example 10 (Normal Distribution) For $X \sim N(\mu, \sigma^2)$,

$$\begin{aligned}\Psi_X(t) &= \int_{-\infty}^{\infty} e^{itx} \frac{1}{\sqrt{2\pi\sigma^2}} \exp\left(-\frac{(x-\mu)^2}{2\sigma^2}\right) dx \\ &= \exp\left(i\mu t - \frac{\sigma^2 t^2}{2}\right)\end{aligned} \tag{2.25}$$

(using completing-the-square and (2.5)). This satisfies (2.21)–(2.24) (Problem 8). ■

Characteristic functions are in one-to-one correspondence with distributions. Moreover, for independent $X_1, \ldots, X_n$,

$$\Psi_{X_1,\ldots,X_n}(t_1, \ldots, t_n) = \Psi_{X_1}(t_1) \cdots \Psi_{X_n}(t_n). \tag{2.26}$$

Example 11 (Binomial Distribution) If $X \in \{0, 1\}$ with $P(X = 1) = p$, then $\Psi_X(t) = 1 - p + pe^{it}$. Using (2.26), the characteristic function of the binomial variable Y in Example 1 is $\Psi_Y(t) = (1 - p + pe^{it})^n$, satisfying (2.21)–(2.24) (Problem 8). This also implies mean np and variance $np(1-p)$. ■

For random variables $X_1, X_2, \ldots$ and cdf $F_X(x) := P(X \leq x)$, we say X_n *converges in law* to X if $F_{X_n}(x_0) \to F_X(x_0)$ at all continuity points x_0 of F_X.

Example 12 (Central Limit Theorem) For X with mean zero, variance σ^2, $\Psi_X(t) = \exp(-\sigma^2t^2/2 + o(t^2))$ near $t = 0$.[5] Let $X_1, \ldots, X_n$ be independent with mean μ and variance σ^2. Then

$$U_n := \frac{X_1 + \cdots + X_n - n\mu}{\sqrt{n}\sigma}$$

[5] $o(t^2)$ means $f(t)/t^2 \to 0$ as $t \to 0$.

has characteristic function

$$\Psi_{U_n}(t) = \left(1 - \frac{t^2}{2n} + o\left(\frac{1}{n}\right)\right)^n \to e^{-t^2/2},$$

the standard normal characteristic function; hence, U_n converges in law to $N(0, 1)$.

2.3 Inverse Wishart Distribution

A distribution is said to have the *reproductive property* if sums of independent random variables with that distribution have the same distributional form (possibly with different parameters).

Proposition 2 (Reproductivity of χ^2) *If*

$$X \sim \chi_l^2, \quad Y \sim \chi_m^2, \quad X \perp\!\!\!\perp Y, \quad \textit{then} \quad X + Y \sim \chi_{l+m}^2.$$

Proof Using $\Psi_X(t) = (1 - 2it)^{-n/2}$ for $X \sim \chi_n^2$ (Problem 9) and independence,

$$\Psi_{X+Y}(t) = (1 - 2it)^{-l/2}(1 - 2it)^{-m/2} = (1 - 2it)^{-(l+m)/2}.$$

By uniqueness of characteristic functions, the claim follows.

■

Similarly, normals are reproductive.

Proposition 3 (Reproductivity of Normals) *If*

$$X \sim N(\mu_X, \sigma_X^2), \quad Y \sim N(\mu_Y, \sigma_Y^2), \quad X \perp\!\!\!\perp Y,$$

then $X + Y \sim N(\mu_X + \mu_Y,\ \sigma_X^2 + \sigma_Y^2)$.

Proof By (2.25) and independence,

$$\Psi_{X+Y}(t) = \mathbb{E}[e^{iXt}]\mathbb{E}[e^{iYt}] = \exp\left(i\mu_X t - \frac{\sigma_X^2 t^2}{2}\right)\exp\left(i\mu_Y t - \frac{\sigma_Y^2 t^2}{2}\right)$$

$$= \exp\left(i(\mu_X + \mu_Y)t - \frac{(\sigma_X^2 + \sigma_Y^2)t^2}{2}\right).$$

Uniqueness yields the result.

■

Let $X_1, \ldots, X_n \sim N(0, \Sigma)$ be independent p-variate normals ($0 \in \mathbb{R}^p$, $\Sigma \in \mathbb{R}^{p\times p}$). Define, for $a > 0$, the p-dimensional Gamma density

$$f_{p,a}(B) := \frac{1}{\Gamma_p(a)} (\det B)^{a-\frac{p+1}{2}} \exp\{-\mathrm{tr}(B)\}, \quad B > 0, \tag{2.27}$$

where B ranges over positive definite symmetric matrices. Here,

$$\Gamma_p(a) = \int_{B>0} (\det B)^{a-\frac{p+1}{2}} \exp\{-\mathrm{tr}(B)\}\,(dB) \tag{2.28}$$

is the p-dimensional Gamma function (an extension of 2.7); $(dB) = \prod_{i=1}^{p}\prod_{j=1}^{i} db_{i,j}$ for the lower triangular entries.

For $p = 1$, (2.7) gives $\Gamma_1(a) = \Gamma(a)$. For $p = 2$, one shows

$$\Gamma_2(a) = \sqrt{\pi}\,\Gamma(a)\,\Gamma\left(a - \frac{1}{2}\right) \tag{2.29}$$

(Problem 11). In general:

Proposition 4

$$\Gamma_p(\alpha) = \pi^{p(p-1)/4} \prod_{j=1}^{p} \Gamma\left(\alpha + \frac{1-j}{2}\right), \qquad \alpha > \frac{p-1}{2}.$$

Proof is given in the appendix.

Next, put $a = \nu/2$, let A be the lower triangular Cholesky factor of Σ ($\Sigma = AA^\top$) with positive diagonal, set $S := (\sqrt{2}A)^{-1}$, and change variables $B = SMS^\top$. This yields the density on positive definite M:

$$f_{\Sigma,\nu}(M) = \frac{(\det \Sigma)^{-\nu/2}}{2^{\nu p/2}\Gamma_p(\nu/2)} (\det M)^{(\nu-p-1)/2} \exp\left\{-\frac{1}{2}\mathrm{tr}(\Sigma^{-1}M)\right\}. \tag{2.30}$$

When $\nu = n \geq p$ is an integer, (2.30) is the density of $M = X_1X_1^\top + \cdots + X_nX_n^\top$, called the *Wishart distribution* [1]. We write $M \sim W(\Sigma, \nu)$, allowing real $\nu > p-1$.

The distribution of $U := M^{-1}$ is the *inverse Wishart*; with $\Lambda := \Sigma^{-1}$, write $U \sim W^{-1}(\Lambda, \nu)$, having density

$$f_{\nu,\Lambda}(U) = \frac{(\det \Lambda)^{\nu/2}}{2^{\nu p/2}\Gamma_p(\nu/2)} (\det U)^{-(\nu+p+1)/2} \exp\left\{-\frac{1}{2}\mathrm{tr}(\Lambda U^{-1})\right\}. \tag{2.31}$$

2.4 Maximum Likelihood Estimation

When a normal distribution is assumed, the mean $\mu \in \mathbb{R}$ and variance $\sigma^2 > 0$ may be known or unknown. In the latter case, from observations $X_1, \ldots, X_n \sim N(\mu, \sigma^2)$, we often estimate

$$\bar{X} := \frac{1}{n}\sum_{i=1}^{n} X_i$$

for μ and

$$S^2 := \frac{1}{n}\sum_{i=1}^{n}(X_i - \bar{X})^2$$

for σ^2. Many estimation methods exist. Here, we explain *maximum likelihood.*

Given $X_1 = x_1, \ldots, X_n = x_n$, the likelihood is $p(x_1, \ldots, x_n \mid \theta)$. If the X_i are independent, this factors as $\prod_{i=1}^{n} p(x_i \mid \theta)$. Viewing it as a function of θ, a maximizer $\hat{\theta} = \hat{\theta}(x_1, \ldots, x_n)$ is the *maximum likelihood estimate* (MLE).

Example 13 Two baseball teams play three games (no ties). There are $2^3 = 8$ possible sequences. Let team A win each game with probability θ (team B with $1 - \theta$). For game i, set $X_i = 1$ if A wins else 0. Assuming independence, for (x_1, x_2, x_3),

$$p(x_1x_2x_3 \mid \theta) = \theta^k(1-\theta)^{3-k},$$

with $k = x_1 + x_2 + x_3$. For $(1, 1, 0)$, $p(110 \mid \theta) = \theta^2(1-\theta)$ (Table 2.1, column 2). Maximizing $p(1, 1, 0 \mid \theta)$ is equivalent to maximizing $L = \log\{\theta^2(1-\theta)\}$:

$$\frac{dL}{d\theta} = \frac{2}{\theta} - \frac{1}{1-\theta} = 0 \iff \theta = \frac{2}{3}.$$

Hence, $\hat{\theta} = 2/3$ and $p(1, 1, 0 \mid \hat{\theta}) = \frac{4}{27}$. ■

Below we assume n independent and identically distributed (I.I.D.) observations from the same distribution.

Example 14 Let X take values in $\{1, \ldots, \alpha\}$. If in $X_1 = x_1, \ldots, X_n = x_n$ and the counts of $1, \ldots, \alpha$ are $k_1, \ldots, k_\alpha$, the likelihood

$$\prod_{i=1}^{n} p(x_i \mid \theta) = \theta_1^{k_1} \cdots \theta_\alpha^{k_\alpha}$$

Table 2.1 Likelihood and maximum likelihood estimator

(x_1, x_2, x_3)	$p(x_1x_2x_3 \mid \theta)$	$p(x_1x_2x_3 \mid \theta = 2/3)$	$\hat{\theta}(x_1, x_2, x_3)$	$p(x_1x_2x_3 \mid \hat{\theta}(x_1, x_2, x_3))$
(0, 0, 0)	$(1-\theta)^3$	1/27	0	1
(1, 0, 0)	$(1-\theta)^2\theta$	2/27	1/3	8/27
(0, 1, 0)	$(1-\theta)^2\theta$	2/27	1/3	8/27
(0, 0, 1)	$(1-\theta)^2\theta$	2/27	1/3	8/27
(0, 1, 1)	$(1-\theta)\theta^2$	4/27	2/3	8/27
(1, 0, 1)	$(1-\theta)\theta^2$	4/27	2/3	8/27
(1, 1, 0)	$(1-\theta)\theta^2$	4/27	2/3	8/27
(1, 1, 1)	θ^3	8/27	1	1

is maximized at the MLEs

$$\hat{\theta}_1 = \frac{k_1}{n}, \ \ldots, \ \hat{\theta}_\alpha = \frac{k_\alpha}{n}$$

(Problem 12). ■

For $\epsilon > 0$, we say X_n *converges in probability* to a if $P(|X_n - a| < \epsilon) \to 1$. In Example 14, $\hat{\theta}_j = k_j/n \to \theta_j$ in probability. This is the *law of large numbers*.

Example 15 Given n data pairs $(x_i, y_i) \in \mathbb{R}^2$, fit a line $y = \beta_0 + \beta_1 x$. Assume $Y \mid (X = x) \sim N(\beta_0 + \beta_1 x, \sigma^2)$. The likelihood is

$$\prod_{i=1}^{n} p(y_i \mid x_i, \beta_0, \beta_1, \sigma^2) = (2\pi\sigma^2)^{-n/2} \exp\left\{-\frac{1}{2\sigma^2}\sum_{i=1}^{n}(y_i - \beta_0 - x_i\beta_1)^2\right\},$$

which is

$$L = -\frac{n}{2}\log(2\pi\sigma^2) - \frac{1}{2\sigma^2}\sum_{i=1}^{n}(y_i - \beta_0 - x_i\beta_1)^2.$$

Differentiating w.r.t. $\beta_0, \beta_1, \sigma^2$ yields three equations. Solving the first two (equivalently least squares) gives $\hat{\beta}_0, \hat{\beta}_1$ (Problem 13); substituting into the third gives

$$\hat{\sigma}^2 = \frac{1}{n}\sum_{i=1}^{n}(y_i - \hat{\beta}_0 - x_i\hat{\beta}_1)^2. \tag{2.32}$$

Plugging back into L yields

$$L = -\frac{n}{2}\log(2\pi\hat{\sigma}^2 e) \tag{2.33}$$

(Problem 14). ■

2.5 Statistical Testing

A teacher claims: "Boys grow 5 cm from the first to the second year of middle school." Suppose growth is normal with known variance $\sigma^2 = 9$ and unknown mean μ. From n observed growth values $x_1, \ldots, x_n$, we will *statistically test* this claim.

Note that we are not asking whether $\overline{x} = \frac{1}{n}\sum x_i$ equals μ. Since there are many boys nationwide, we treat $x_1, \ldots, x_n$ as one random sample from $N(\mu, \sigma^2)$, tolerating some error in estimation/testing.

We take the hypothesis $\mu = 5$ as the *null hypothesis*. If assuming it leads to a contradiction with data, we reject it; otherwise we do not reject it. The competing claim $\mu \neq 5$ is the *alternative hypothesis*. Testing assesses whether the data provides sufficient evidence for the alternative.

There are two kinds of testing errors:

Type I error Reject the null though it is true.
Type II error Fail to reject the null though it is false.

Examples include medical diagnosis (healthy vs. ill), quality control (good vs. defective), and policing (innocent vs. guilty).

We set a significance level α (e.g., 5%) for type I error and aim to minimize β (the type II error rate) or equivalently maximize the *power* $1 - \beta$. Multiple testing procedures may be available; for fixed α, a test with larger power $1-\beta$ is preferable. We can visualize performance via the *ROC curve*; a larger Area Under Curve (AUC) indicates a better test.

Tests on the same ROC curve use the same method but different thresholds (e.g., fever threshold 37°C vs. 38°C). Different methods (e.g., a saliva-based assay) yield different ROC curves; comparing AUCs helps select the better method (Fig. 2.2).

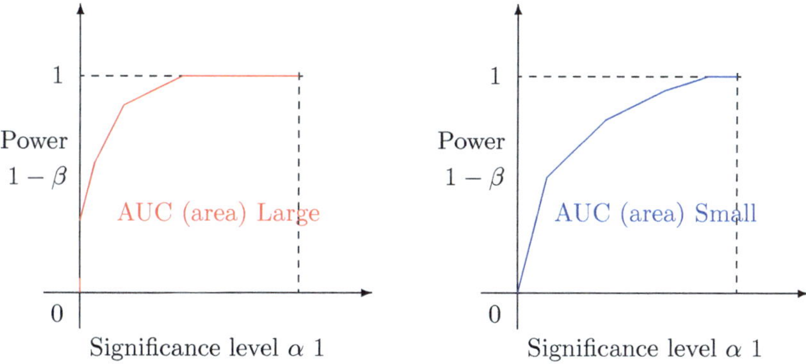

Fig. 2.2 ROC curves: a test with a large AUC vs. a test with a small AUC

Fixing α means we tolerate type I errors up to probability α. If $X_i \sim N(\mu, \sigma^2)$, then

$$\frac{1}{n}\sum_{i=1}^{n} X_i \sim N\left(\mu, \frac{\sigma^2}{n}\right).$$

Indeed,

$$\mathbb{E}\left[\frac{1}{n}\sum_{i=1}^{n} X_i\right] = \mu,$$

$$\mathbb{V}(\overline{X}) = \frac{1}{n^2}\sum_{i=1}^{n}\sum_{j=1}^{n} \operatorname{cov}(X_i, X_j) = \frac{1}{n^2}\cdot n\sigma^2 = \frac{\sigma^2}{n},$$

using independence for $i \neq j$. Hence,

$$\frac{\sqrt{n}(\overline{X} - \mu)}{\sigma} \sim N(0, 1) \tag{2.34}$$

(Problem 15).

A standard two-sided test of $H_0 : \mu = 5$ uses

$$Z := \frac{\sqrt{n}(\overline{X} - 5)}{3},$$

which is standard normal under H_0. We reject H_0 if Z falls in the two tails with probability $\alpha/2$ each. The distribution of the test statistic under H_0 is the *null distribution* (Fig. 2.3).

In our example, a two-sided test takes $H_0 : \mu = 5$ vs. $H_1 : \mu \neq 5$, placing rejection regions of size $\alpha/2$ in both tails (Figure 2.4a). One-sided tests use a single tail: $H_1 : \mu > 5$ (right-sided, Fig. 2.4c) or $H_1 : \mu < 5$ (left-sided, Fig. 2.4b). In Chap. 4, tests of (conditional) independence are one-sided.

Example 16 For $n = 10$ growth values

```
5.0, 4.9, 5.3, 5.2, 3.9, 4.5, 6.1, 4.8, 5.2, 4.3
```

and $\alpha = 0.05$ two-sided, the following R code shows H_0 is not rejected:

```
x <- c(5.0, 4.9, 5.3, 5.2, 3.9, 4.5, 6.1, 4.8, 5.2, 4.3)
mu <- 5
sigma <- 3
n <- length(x)
sqrt(n) * (mean(x)-mu) / sigma    ## value of Z
qnorm(0.025)                      ## lower critical value
qnorm(0.975)                      ## upper critical value
```

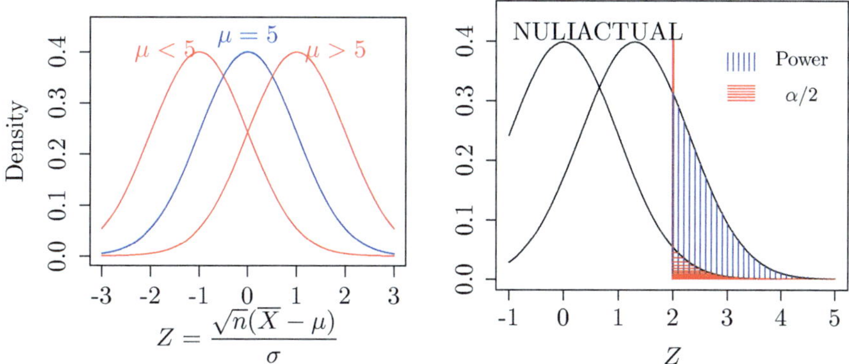

Fig. 2.3 The distribution of Z varies with the true mean (left). Under the null, Z in the rejection region (right tail) constitutes a type I error; under the true (shifted) distribution, the blue vertical area indicates the power

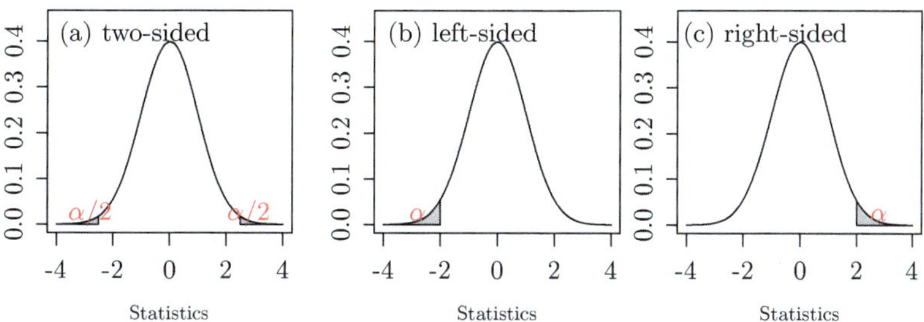

Fig. 2.4 Two-sided and one-sided tests

```
> sqrt(n)*(mean(x)-mu)/sigma
[1] -0.0843274
> qnorm(0.025)
[1] -1.959964
> qnorm(0.975)
[1] 1.959964
```

■

Appendix

We identify a positive definite symmetric matrix or a lower triangular matrix with the vector consisting of its $p(p+1)/2$ lower triangular elements. The Jacobians of the following transformations are known. For Wishart and inverse Wishart results and the proof of Proposition 5, see Chapter 2 of [18].

Proposition 5 *For a positive definite symmetric,* $X \in \mathbb{R}^{p\times p}$*:*

1. *For non-singular* $S \in \mathbb{R}^{p\times p}$*, with* $X = SYS^\top$ *and* $T = S^\top S$*,*

$$(dX) = (\det T)^{\frac{p+1}{2}}(dY), \quad \det X = (\det T)(\det Y), \quad \mathrm{tr}(X) = \mathrm{tr}(TY).$$

2. *For* $X = Y^{-1}$*,* $(dX) = (\det Y)^{-(p+1)}(dY)$*.*
3. *For* $X = YY^\top$ *with lower triangular* $Y = (y_{ij})$ *having positive diagonal,*

$$(dX) = 2^p \prod_{i=1}^{p} y_{ii}^{p+1-i} \cdot \prod_{i=1}^{p}\prod_{j=1}^{i} dy_{ij}, \quad \det X = \prod_{i=1}^{p} y_{ii}^2, \quad \mathrm{tr}(X) = \sum_{i=1}^{p}\sum_{j=1}^{i} y_{ij}^2.$$

Using Proposition 5, we prove (2.30), (2.31), and Proposition 4, in that order.

Proofs of (2.30) and (2.31)

From Proposition 5(1), with $T = (2\Sigma)^{-1}$ and $Y = M$,

$$\begin{aligned}(dB) &= (\det(2\Sigma)^{-1})^{\frac{p+1}{2}}(dM) = 2^{-\frac{p(p+1)}{2}}(\det\Sigma)^{-\frac{p+1}{2}}(dM),\\ \det(B) &= 2^{-p}(\det\Sigma)^{-1}\det(M),\\ \mathrm{tr}(B) &= \tfrac{1}{2}\,\mathrm{tr}(\Sigma^{-1}M).\end{aligned}$$

Substituting into (2.27) gives (2.30).

For (2.31), by Proposition 5(2), $(dM) = (\det U)^{-(p+1)}(dU)$. Substitute $M = U^{-1}$ into (2.30) and multiply by $(\det U)^{-(p+1)}$.

Proof of Proposition 4

Via Cholesky, any $B > 0$ can be written $B = UU^\top$ with lower triangular U having positive diagonal. By Proposition 5(3),

$$\begin{aligned}\Gamma_p(a) &= 2^p \int\cdots\int \exp\left(-\sum_{i=1}^{p}\sum_{j=1}^{i} u_{ij}^2\right)\left(\prod_{i=1}^{p} u_{ii}^2\right)^{a-(p+1)/2} \prod_{i=1}^{p} u_{ii}^{p+1-i} \prod_{i=1}^{p}\prod_{j=1}^{i} du_{ij}\\ &= \left\{\int_{-\infty}^{\infty} e^{-s^2} ds\right\}^{p(p-1)/2} \prod_{i=1}^{p}\left\{\int_0^{\infty} e^{-u_{ii}^2}(u_{ii}^2)^{a-i/2}\, 2\, du_{ii}\right\}\end{aligned}$$

$$= \pi^{p(p-1)/4} \prod_{i=1}^{p} \Gamma\left(a - \frac{i-1}{2}\right),$$

which is the claimed formula (Problem 17). Here, B-integration corresponds to integrating over $u_{ii} \geq 0$ and $u_{ij} \in \mathbb{R}$ for $i \neq j$, with $(dB) = \prod_{i=1}^{p} \prod_{j=1}^{i} du_{ij}$. The last step uses the normal integral with $\sigma^2 = 1/2$ and the Gamma density (2.9) with a replaced by $a - (p-1)/2$. ∎

Problems 1–18

1. Derive (2.1) and (2.2) from the three axioms of probability. [Hint] Regarding (2.2), note that

$$A \cup B = (A - A \cap B) \cup (B - A \cap B) \cup A \cap B,$$
$$A = (A - A \cap B) \cup A \cap B,$$
$$B = (B - A \cap B) \cup A \cap B$$

where in each case, the event on the left-hand side is written as the union of disjoint events on the right-hand side, so each term can be replaced by its probability. From this, the desired relation follows.
2. Show the last two equalities of (2.6).
3. Show (2.8) and $\Gamma(1/2) = \sqrt{\pi}$. [Hint] In (2.7) with $\alpha = 1/2$, perform the substitution $t = x^2/2$ in the integral.
4. Prove Eqs. (2.12), (2.13), (2.14), and (2.19). You may assume continuous random variables.
5. For discrete random variables, show that the validity of (2.16) for all $x, y \in \mathbb{N}$ is equivalent to the validity of (2.17) for all $a, b \in \mathbb{N}$. [Hint] Since X and Y take at most countably many values, we may assume they take values in the nonnegative integers. Using

$$P(X \leq x, Y \leq y) = \sum_{a=0}^{x} \sum_{b=0}^{y} P(X = a, Y = b),$$

one can prove (2.17) $\Rightarrow$ (2.16). For the converse, substitute $(x, y) = (a, b), (a, b-1)$ into (2.16) to obtain a relation. Replace a by $a - 1$ to obtain another relation, and combine them to obtain the final relation.
6. Assuming $\mathbb{V}[Y] > 0$, let $Z := X - \dfrac{cov(X, Y)}{\mathbb{V}[Y]} Y$. Show that $cov(Y, Z) = 0$ and that

$$\mathbb{V}[X] \geq \frac{cov(X, Y)^2}{\mathbb{V}[Y]}.$$

[Hint] Substitute $X = Z + \frac{cov(X,Y)}{\mathbb{V}[Y]} Y$ into $\mathbb{V}[X]$.

7. Show (2.20). [Hint] Transform as follows:

$$\frac{1}{2\pi\sigma_X\sigma_Y\sqrt{1-\rho^2}}$$
$$\cdot \exp\left\{-\frac{1}{2(1-\rho^2)}\left[\left(\frac{x-\mu_X}{\sigma_X}\right)^2 - 2\rho\left(\frac{x-\mu_X}{\sigma_X}\right)\left(\frac{y-\mu_Y}{\sigma_Y}\right)\right.\right.$$
$$\left.\left.+\left(\frac{y-\mu_Y}{\sigma_Y}\right)^2\right]\right\}$$
$$= \frac{1}{\sqrt{2\pi\sigma_X^2}} \exp\left\{-\frac{1}{2}\left(\frac{x-\mu_X}{\sigma_X}\right)^2\right\}$$
$$\cdot \frac{1}{\sqrt{2\pi\sigma_Y^2(1-\rho^2)}} \exp\left\{-\frac{1}{2(1-\rho^2)}\left(\frac{y-\mu_Y}{\sigma_Y} - \rho\frac{x-\mu_X}{\sigma_X}\right)^2\right\}.$$

8. In Examples 10 and 11, show that the obtained characteristic functions satisfy (2.21)–(2.24). [Hint] For Example 11, use

$$\Psi_X{}'(t) = inp(1-p+pe^{it})^{n-1}e^{it},$$
$$\Psi_X{}''(t) = -npe^{it}\left(1-p+pe^{it}\right)^{n-1} - np^2(n-1)e^{2it}\left(1-p+pe^{it}\right)^{n-2}.$$

9. For $X \sim \chi_n^2$, show that $\Psi_X(t) = (1-2it)^{-n/2}$. [Hint] First derive

$$\Psi_X(t) = \frac{1}{2^{n/2}\Gamma(n/2)} \int_0^\infty x^{n/2-1} \exp\left(-\frac{1-2it}{2}x\right) dx,$$

and then let $y := (1-2it)x$.

10. Based on the central limit theorem, suppose you wish to generate a standard normal random variable by tossing a coin n times (with n sufficiently large). How can you construct such a random variable from the coin toss results? Also, generate such random variables 1000 times and output a histogram of their frequencies in R. [Hint] Let $X_i = 1$ if heads and $X_i = 0$ if tails. Then $S_n = \sum_{i=1}^n X_i$ has mean $n/2$ and variance $n/4$.

11. Consider the derivation of (2.29).
    (a) Derive

$$J=\det\begin{bmatrix} \frac{\partial b_{11}}{\partial u} & \frac{\partial b_{11}}{\partial v} & \frac{\partial b_{11}}{\partial \theta} \\ \frac{\partial b_{22}}{\partial u} & \frac{\partial b_{22}}{\partial v} & \frac{\partial b_{22}}{\partial \theta} \\ \frac{\partial b_{12}}{\partial u} & \frac{\partial b_{12}}{\partial v} & \frac{\partial b_{12}}{\partial \theta} \end{bmatrix} = \det\begin{bmatrix} 1 & 0 & 0 \\ 0 & 1 & 0 \\ \frac{\partial b_{12}}{\partial u} & \frac{\partial b_{12}}{\partial v} & \sqrt{uv}\sin\theta \end{bmatrix} = \sqrt{uv}\sin\theta .$$

    (b) Show

$$\int_0^{\pi} \sin^{2(a-1)}\theta d\theta = \frac{\Gamma(a-\frac{1}{2})\Gamma(\frac{1}{2})}{\Gamma(a)}.$$

    [Hint] Set $t=\cos^2\theta$ and use substitution. Since the integrand is even, halve the interval. Note that $\frac{dt}{d\theta}=-2\cos\theta\sin\theta=-2(1-t)^{1/2}t^{1/2}$. Then apply (1.11).
    (c) Show (2.29).

12. Derive the maximum likelihood estimator in Example 14. [Hint] The MLE maximizes ℓ under the constraint $\sum p_i = 1$. Consider

$$\sum_{i=1}^{\alpha} k_i \log p_i + \lambda\left(1-\sum_{i=1}^{\alpha} p_i\right),$$

differentiate with respect to p_j, set equal to 0, express p_j in terms of λ, and use $\sum p_i = 1$ to solve for λ.

13. In Example 15, show that the solutions for $\hat{\beta}_0, \hat{\beta}_1$ are

$$\hat{\beta}_1 = \frac{\sum_{i=1}^{n}(x_i-\bar{x})(y_i-\bar{y})}{\sum_{i=1}^{n}(x_i-\bar{x})^2}, \qquad \hat{\beta}_0 = \bar{y}-\bar{x}\hat{\beta}_1 .$$

First solve the simultaneous equations $\frac{\partial L}{\partial \beta_0}=0$, $\frac{\partial L}{\partial \beta_1}=0$. Begin by assuming $\bar{x}=\bar{y}=0$, and then replace (x_i, y_i) by $(x_i-\bar{x}, y_i-\bar{y})$ to show that the slope does not change. Use this to obtain $\hat{\beta}_1$, and then determine $\hat{\beta}_0$ from $\bar{x}, \bar{y}, \hat{\beta}_1$.

14. Show (2.32) and (2.33).
15. Prove (2.34).

16. Regarding Example 16:

    (a) Generate height data for ten junior high school students from $N(5, 3^2)$, set $\alpha = 0.05$, and perform a statistical test using R.
    (b) Generate height data for ten junior high school students from $N(7, 3^2)$, and check whether the test statistic Z falls in the rejection region. Repeat this operation 1000 times for $\alpha = 0.05$ and estimate the detection rate. Also perform the same for $\alpha = 0.01$.

17. In the proof of Proposition 4 in the appendix, explain why the five equalities in the derivation of $\Gamma_p(a)$ hold.
18. Let Y_p be a lower triangular matrix with all positive diagonal entries and $X_p = Y_p Y_p^\top$. Show that

$$X_p = \begin{bmatrix} Y_{p-1} & 0 \\ y_{p-1} & y_{pp} \end{bmatrix} \begin{bmatrix} Y_{p-1}^\top & y_{p-1}^\top \\ 0 & y_{pp} \end{bmatrix} = \begin{bmatrix} X_{p-1} & Y_{p-1} y_{p-1}^\top \\ y_{p-1} Y_{p-1}^\top & y_{p-1} y_{p-1}^\top + y_{pp}^2 \end{bmatrix}.$$

Using this, prove part 3 of Proposition 5.

   (a) Show that the Jacobian matrix of $\tilde{X}_p$ (consisting of the lower triangular entries of X_{p-1}, $y_{p-1}Y_{p-1}^\top$, and $y_{p-1}y_{p-1}^\top + y_{pp}^2$) with respect to $\tilde{Y}_p$ (the lower triangular entries of Y_p) can be written as

$$\frac{\partial \tilde{X}_p}{\partial \tilde{Y}_p} = \begin{bmatrix} \dfrac{\partial \tilde{X}_{p-1}}{\partial \tilde{Y}_{p-1}} & * & * \\ 0 & Y_{p-1}^\top & * \\ 0 & 0 & 2y_{p,p} \end{bmatrix}.$$

   (b) Compute $\dfrac{\partial \tilde{X}_1}{\partial \tilde{Y}_1}$. Assuming that $\det\left(\dfrac{\partial \tilde{X}_{p-1}}{\partial \tilde{Y}_{p-1}}\right) = 2^{p-1}\displaystyle\prod_{i=1}^{p-1} y_{ii}^{p-i}$ holds, show that $\det\left(\dfrac{\partial \tilde{X}_p}{\partial \tilde{Y}_p}\right) = 2^p \displaystyle\prod_{i=1}^{p} y_{ii}^{p+1-i}$. [Hint] Use $\det(Y_{p-1}) = \prod_{i=1}^{p-1} y_{ii}$.

   (c) Find X_1. Then, assuming that $\mathrm{tr}(X_{p-1}) = \displaystyle\sum_{i=1}^{p-1}\sum_{j=1}^{i} y_{i,j}^2$ holds, show that $\mathrm{tr}(X_p) = \displaystyle\sum_{i=1}^{p}\sum_{j=1}^{i} y_{i,j}^2$. [Hint] Use $y_{p-1} = [y_{p,1}, \ldots, y_{p,p-1}]$.

Chapter 3
Graphical Models

In this chapter, we study graphical models, which express conditional independence among random variables. A graphical model is a framework that visually represents probabilistic structure using either undirected graphs or directed acyclic graphs (DAGs).

We begin by introducing the graphoid axioms—symmetry, decomposition, weak union, and contraction—as the basic properties that conditional independence should satisfy and explain their meanings and roles. Next, we define separation in undirected graphs and d-separation in DAGs and formalize how these correspond to conditional independence.

We then define Markov networks (MNs) based on undirected graphs and Bayesian networks based on DAGs, illustrating the differences and characteristics of the probabilistic structures they can represent. Finally, we introduce the Chow–Liu algorithm and demonstrate its usefulness as a method for constructing approximate distributions using mutual information. By using this algorithm to construct a maximum spanning tree of an undirected graph, we obtain an approximation that is optimal in terms of Kullback–Leibler divergence.

3.1 Conditional Independence

As an extension of independence, we have *conditional independence*.

When a random variable Z takes the value $Z = z \in \mathbb{R}$, we say random variables X and Y are independent given $Z = z$ if, for any $x, y \in \mathbb{R}$,

$$P(X \le x, Y \le y \mid Z = z) = P(X \le x \mid Z = z)\, P(Y \le y \mid Z = z).$$

If this holds for $z \in \mathbb{R}$ with probability one, we say that X and Y are conditionally independent given Z, and write $X \perp\!\!\!\perp Y \mid Z$ (if X and Y are simply independent, we write $X \perp\!\!\!\perp Y$).

J. Suzuki, *Graphical Models and Causal Discovery with R*,
https://doi.org/10.1007/978-981-95-4267-3_3

Accordingly, in the discrete and continuous cases we have, with probability one,

$$P(X = x, Y = y \mid Z = z) = P(X = x \mid Z = z)\, P(Y = y \mid Z = z),$$

$$f_{XY|Z}(x, y \mid z) = f_{X|Z}(x \mid z)\, f_{Y|Z}(y \mid z),$$

where $f_{XYZ}(x, y, z)$ denotes the joint density of X, Y, Z, and

$$f_{XY|Z}(x, y, z) := \frac{f_{XYZ}(x, y, z)}{f_Z(z)}.$$

Equivalently, one may write

$$P(X = x, Y = y, Z = z)\, P(Z = z) = P(X = x, Z = z)\, P(Y = y, Z = z),$$

$$f_{XYZ}(x, y, z)\, f_Z(z) = f_{XZ}(x, z)\, f_{YZ}(y, z).$$

Conditional independence can also be defined for multiple random variables, e.g.,

$$\{X, Y\} \perp\!\!\!\perp Z \mid \{U, V\}, \tag{3.1}$$

meaning "$\{X, Y\}$ and Z are conditionally independent given $\{U, V\}$." When representing multiple variables such as X, Y or U, V, we enclose them in braces $\{\cdot\}$.

Below, we sometimes suppress explicit mention of the argument, writing $P(X)$, f_X instead of $P(X = x)$, $f_X(x)$ when the value is clear from context.

Example 17 (Conditional Independence) Let the sample space be $\Omega = \{1, 2, 3, 4, 5, 6\}$, and consider events $A = \{1\}$, $B = \{1, 2\}$, $C = \{1, 2, 3\}$. Since

$$P(A \cap C \mid B) = P(A \mid B)\, P(C \mid B) = P(\{1\} \mid \{1, 2\}),$$

the events A and C are conditionally independent given B.

Example 18 (Conditional Independence and Its Graphical Representation) For discrete random variables X, Y, Z, if the joint distribution can be written as

$$P(X, Y, Z) = \frac{P(X, Z)\, P(Y, Z)}{P(Z)}, \tag{3.2}$$

then $X \perp\!\!\!\perp Y \mid Z$ holds. Indeed, for a specific value $Z = z$,

$$P(X, Y \mid Z = z) = P(X \mid Z = z)\, P(Y \mid Z = z),$$

so X and Y are independent given $Z = z$.

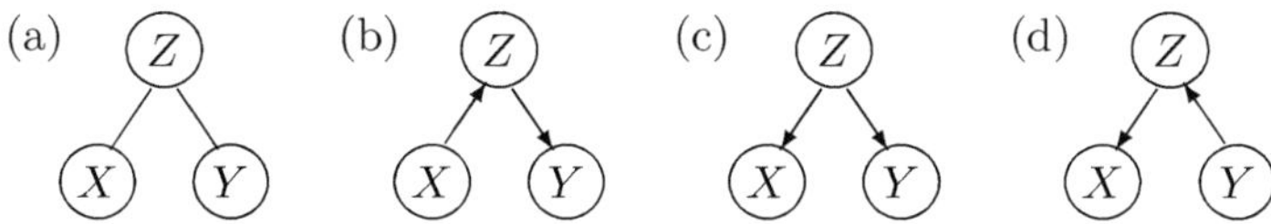

Fig. 3.1 The conditional independence $X \perp\!\!\!\perp Y \mid Z$ represented by the undirected graph (**a**) and the directed graphs (**b**), (**c**), and (**d**)

Equation (3.2) can be expressed in three ways:

$$P(X)\,P(Z \mid X)\,P(Y \mid Z) = P(Z)\,P(X \mid Z)\,P(Y \mid Z) = P(Y)\,P(Z \mid Y)\,P(X \mid Z). \tag{3.3}$$

These three factorizations are often represented by the undirected graph in Fig. 3.1a and the directed (acyclic) graphs in Fig. 3.1b, c, and d. ■

For conditional independence, the following proposition holds. These are the "graphoid axioms," important properties in the analysis of graphical models and cited later in the book.

Proposition 6 (Pearl–Paz [20]) *Let A, B, C, D be pairwise disjoint subsets of $\{1, \ldots, p\}$. Then*

$$X_A \perp\!\!\!\perp X_B \mid X_C \iff X_B \perp\!\!\!\perp X_A \mid X_C, \tag{3.4}$$

$$X_A \perp\!\!\!\perp X_{B\cup D} \mid X_C \Longrightarrow X_A \perp\!\!\!\perp X_B \mid X_C \text{ and } X_A \perp\!\!\!\perp X_D \mid X_C, \tag{3.5}$$

$$X_A \perp\!\!\!\perp X_{B\cup D} \mid X_C \Longrightarrow X_A \perp\!\!\!\perp X_B \mid X_{C\cup D}, \tag{3.6}$$

$$X_A \perp\!\!\!\perp X_B \mid X_C \text{ and } X_A \perp\!\!\!\perp X_D \mid X_{B\cup C} \Longrightarrow X_A \perp\!\!\!\perp X_{B\cup D} \mid X_C \tag{3.7}$$

hold.

In Proposition 6, if $A = \{1, 2\}$ and $B = \{2, 4\}$ then $X_A = \{X_1, X_2\}$ and $X_{A\cup B} = \{X_1, X_2, X_4\}$. The properties (3.4)–(3.7) are called *symmetry*, *decomposition*, *weak union*, and *contraction*, respectively.

Proof Symmetry follows from

$$P(X_A, X_B \mid X_C) = P(X_A \mid X_C)P(X_B \mid X_C) = P(X_B \mid X_C)P(X_A \mid X_C).$$

For decomposition, marginalizing $X_A \perp\!\!\!\perp X_{B\cup D} \mid X_C$ over X_D yields $X_A \perp\!\!\!\perp X_B \mid X_C$, and marginalizing over X_B yields $X_A \perp\!\!\!\perp X_D \mid X_C$. For weak union, from the assumption, we have $X_A \perp\!\!\!\perp X_{B\cup D} \mid X_C$ and by decomposition also $X_A \perp\!\!\!\perp X_D \mid X_C$. Hence,

$$P(X_A \mid X_C) = P(X_A \mid X_{B\cup C\cup D}), \qquad P(X_A \mid X_C) = P(X_A \mid X_{C\cup D}),$$

which implies

$$P(X_A \mid X_{B\cup C\cup D}) = P(X_A \mid X_{C\cup D}),$$

i.e., weak union holds. For contraction, the assumptions

$$X_A \perp\!\!\!\perp X_B \mid X_C, \qquad X_A \perp\!\!\!\perp X_D \mid X_{B\cup C}$$

are equivalent to

$$P(X_A \mid X_C) = P(X_A \mid X_{B\cup C}), \qquad P(X_A \mid X_{B\cup C}) = P(X_A \mid X_{B\cup C\cup D}),$$

and hence

$$P(X_A \mid X_C) = P(X_A \mid X_{B\cup C\cup D}),$$

which proves contraction.

■

3.2 Separation in Graphs

In this book, conditional independence among random variables is represented using undirected graphs and directed acyclic graphs. We introduce the notion of separation needed for this understanding.

Prepare p vertices $1, \ldots, p$ and construct a graph G by connecting them with edges. First, fix the *vertex set* $V = \{1, \ldots, p\}$. We consider undirected graphs consisting only of undirected edges and directed graphs consisting only of directed edges.

In an *undirected graph*, an undirected edge connecting vertices i, j is denoted by the set $\{i, j\}$. Since it is a set, $\{i, j\} = \{j, i\}$. The *edge set* E is defined as a subset of $\mathcal{E} := \{\{i, j\} \mid i, j \in V,\ i \neq j\}$. If $\{i, j\} \in E$, we say there exists a *path* from i to j. If $i, j \in V$ are connected by a path and $\{j, k\} \in E$, then there exists a path from i to k.

Example 19 (Undirected Graphs) For the vertex set $V = \{1, 2, 3\}$, there are eight undirected graphs $G = (V, E)$ as shown in Fig. 3.2. Their edge sets E are

$$\{\},\ \{\{1,2\}\},\ \{\{2,3\}\},\ \{\{3,1\}\},\ \{\{3,1\},\{3,2\}\},$$
$$\{\{1,2\},\{1,3\}\},\ \{\{2,3\},\{2,1\}\},\ \{\{2,3\},\{3,1\},\{1,2\}\}.$$

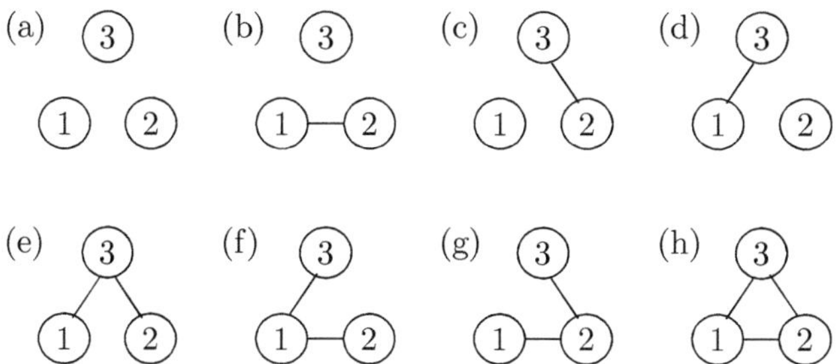

Fig. 3.2 For an undirected graph with three vertices, there are 2^3 possible types depending on the presence or absence of edges

In (e), (f), and (g), there is no path from 1 to 2, from 2 to 3, and from 1 to 3, respectively. In (h), there exists a cycle (a path that starts at a vertex and returns to it). ■

In a *directed graph*, a directed edge from vertex i to j is denoted by the ordered pair (i, j), so $(i, j) \neq (j, i)$. The edge set E is a subset of $\mathcal{E} := \{(i, j) \mid i, j \in V,\ i \neq j\}$. If $(i, j) \in E$, we say there exists a *directed path* from i to j, or that j is a *descendant* of i. If there exists a directed path from i to j and $(j, k) \in E$, then there exists a directed path from i to k. If all vertices on a directed path are distinct, it is called a *directed acyclic path*. If all directed paths are acyclic, the directed graph $G = (V, E)$ is called a *directed acyclic graph (DAG)*. Below, we will refer to directed acyclic graphs as DAGs. A path that ignores the direction of each edge and follows connections is called an *undirected path*.

Example 20 (DAG) As there are 2^3 undirected graphs when $V = \{1, 2, 3\}$, there are $3^3 = 27$ directed graphs on $p = 3$ vertices (for each pair of vertices, there are three possibilities: no edge, an edge in one direction, or in the other). However, two of these, shown in Fig. 3.3, contain directed cycles, so there are 25 DAGs. ■

Next we define separation in graphs.

First, consider, in an undirected graph, the undirected paths (possibly multiple) from i to j for two distinct vertices $i, j \in V$. Let C be a subset of V that does not contain i or j. If every path from i to j passes through at least one vertex in C, we say that i and j are *separated* by C.

If A, B, C are pairwise disjoint subsets of V, and every $i \in A$ and $j \in B$ are separated by C, then we say that A and B are separated by C, and write $A \perp\!\!\!\perp_G B \mid C$.

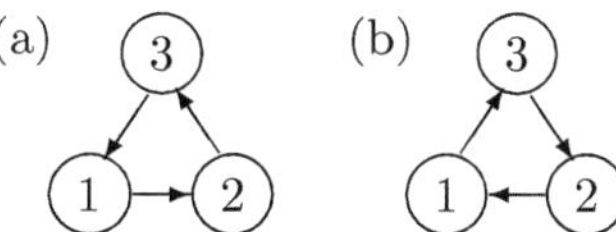

Fig. 3.3 Among the 27 directed graphs with $p = 3$ vertices, cases (**a**) and (**b**) contain directed cycles

Example 21 (Separation in an Undirected Graph) For the undirected graph in Fig. 3.5a, assess the truth of the following:

1. $\{5\} \perp\!\!\!\perp_G \{12\} \mid \{9\}$: There exists a path $5 \to 8 \to 12$ not meeting $\{9\}$; false.
2. $\{7\} \perp\!\!\!\perp_G \{9\}$: There exists a path $7 \to 3 \to 5 \to 9$; false.
3. $\{1\} \perp\!\!\!\perp_G \{10\} \mid \{5, 8\}$: Any path from 1 to 10 must pass through 5 or 8; true.
4. $\{7\} \perp\!\!\!\perp_G \{10\} \mid \{13\}$: Any path from 7 to 10 must pass through 13; true.
5. $\{1\} \perp\!\!\!\perp_G \{12\} \mid \{4, 9, 11\}$: There is a path $1 \to 2 \to 5 \to 8 \to 12$ not meeting $\{4, 9, 11\}$; false.
6. $\{2, 4\} \perp\!\!\!\perp_G \{6, 13\} \mid \{5, 12\}$: Every path connecting 2 to 6, 2 to 13, 4 to 6, and 4 to 13 must pass through 5 or 12; true.

■

Next consider DAGs. For two distinct vertices $i, j \in V$, consider the undirected paths (possibly multiple) between i and j. Let C be a subset of V not containing i or j. If every undirected path from i to j contains a vertex k satisfying one of the following conditions, then we say i and j are *d-separated* by C (see Fig. 3.4).

1. The directions of the two edges adjacent to k along the path are $\to k \to$, $\leftarrow k \leftarrow$, or $\leftarrow k \to$, and $k \in C$.
2. The directions are $\to k \leftarrow$ (a *collider*), and *neither k nor any of its descendants belong to C.*

If A, B, C are pairwise disjoint subsets of V and every $i \in A$ and $j \in B$ are d-separated by C, then we say A and B are d-separated by C, and write $A \perp\!\!\!\perp_G B \mid C$. In this book, the case "$\to k \leftarrow$" is called a *collider (head-to-head).*

Example 22 (D-separation in a DAG) From Fig. 3.5b, assess the truth of the following:

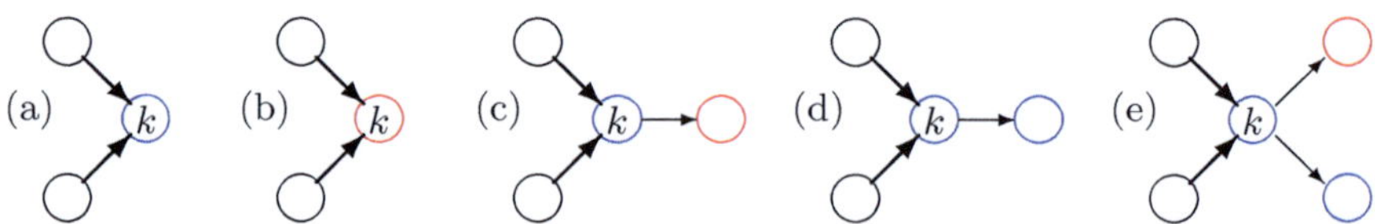

Fig. 3.4 (**a**) When a collision occurs at k on the path, it is separated since $k \notin C$ and has no descendants; (**b**) not separated because $k \in C$; (**c**) not separated because $k \notin C$ but a descendant is in C; (**d**) separated because $k \notin C$ and none of its descendants are in C; (**e**) not separated because $k \notin C$ but a descendant is in C

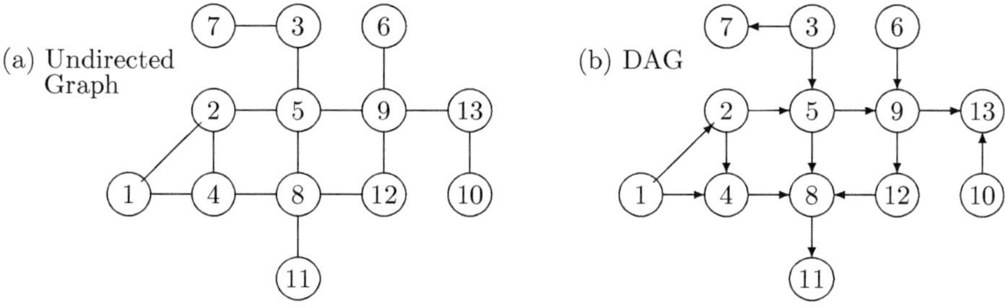

Fig. 3.5 The undirected graph in Example 21 and the DAG in Example 22

1. $\{5\} \perp\!\!\!\perp_G \{12\} \mid \{9\}$: Both $5 \to 9 \to 12$ and $5 \to 8 \to 12$ are blocked; true. Indeed, 8 and its descendant 11 are not in the separating set $\{9\}$, so although $5 \to 8 \leftarrow 12$ is a collider, it remains blocked.
2. $\{7\} \perp\!\!\!\perp_G \{9\}$: There exists a path $7 \to 3 \to 5 \to 9$ that is not blocked; false.
3. $\{1\} \perp\!\!\!\perp_G \{10\} \mid \{5, 8\}$: Every path goes through 13, which is a collider not included in the separating set; thus blocked; true.
4. $\{7\} \perp\!\!\!\perp_G \{10\} \mid \{13\}$: Any path from 7 to 10 must pass through 13, but 13 is a collider and is included in the separating set $\{13\}$, which opens the collider; hence not blocked; false.
5. $\{1\} \perp\!\!\!\perp_G \{12\} \mid \{4, 9, 11\}$: The path $1 \to 2 \to 5 \to 8 \to 12$ has a collider at 8, but its descendant 11 is in the separating set $\{4, 9, 11\}$, so the path is unblocked; false.
6. $\{2, 4\} \perp\!\!\!\perp_G \{6, 13\} \mid \{5, 12\}$: Every path connecting 2 to 6 or 13, and 4 to 6 or 13, must pass through 5 or 12; true.

■

Whether we use an undirected graph or a DAG will be clear from context, so we use the same notation $G = (V, E)$. Separation applies to undirected graphs, and d-separation to DAGs.

The following proposition holds for both separation and d-separation.

Proposition 7 *Let A, B, C, D be pairwise disjoint subsets of $\{1, \ldots, p\}$. Then*

$$A \perp\!\!\!\perp_G C \mid D \text{ and } B \perp\!\!\!\perp_G C \mid D \iff (A \cup B) \perp\!\!\!\perp_G C \mid D, \tag{3.8}$$

Proof See Problem 25.

3.3 Representing Conditional Independence with Graphs

A *graphical model* is the representation of conditional independence among random variables via separation properties in a graph. We identify random variables with vertices.

Let A, B, C be disjoint subsets of the vertex set $V = \{1, 2, \ldots, p\}$, identified with $\{X_1, \ldots, X_p\}$, and choose the edge set E so that

$$A \perp\!\!\!\perp_G B \mid C \Longrightarrow X_A \perp\!\!\!\perp X_B \mid X_C. \tag{3.9}$$

However, the converse

$$A \perp\!\!\!\perp_G B \mid C \Longleftarrow X_A \perp\!\!\!\perp X_B \mid X_C \tag{3.10}$$

need not hold for any E.

If we start from $E = \mathcal{E}$ (all possible edges), then nothing is separated and $A \perp\!\!\!\perp_G B \mid C$ is false, so (3.9) holds trivially. From there, we successively remove edges while maintaining (3.9) and minimize the number of edges (details in Example 25). When G is undirected, such a model is called a *Markov network* (MN), and when G is a DAG, it is called a *Bayesian network* (BN) [14, 20, 31]. In general, even when the set of conditional independencies to be represented is fixed, the edge set E need not be unique. We illustrate this with examples.

Example 23 Equation (3.2) in Example 18 implies $X \perp\!\!\!\perp Y \mid Z$, and Fig. 3.1a is the MN representation, while Fig. 3.1b, c, and d are BN representations. In (a), $X \perp\!\!\!\perp_G Y \mid Z$ holds by (undirected) separation; in (b), (c), and (d), $X \perp\!\!\!\perp_G Y \mid Z$ holds by d-separation. There are no other nontrivial separations or d-separations in these graphs. Thus, in this case, (3.10) also holds. ■

As seen in Example 23, a BN is generally not unique. Figures 3.1b, c, and d correspond to the three factorizations (3.3) and represent the same distribution. BNs that represent the same distribution are said to be *Markov equivalent*.

Example 24 There are 25 DAGs on $p = 3$ vertices (Example 20), but identifying Markov equivalent BNs yields the 11 classes in Fig. 3.6 (see Problem 19). For example,

$$P(X)\,P(Y)\,P(Z),$$

$$P(X)\,P(YZ),\ \ P(Y)\,P(ZX),\ \ P(Z)\,P(XY),$$

$$\frac{P(ZX)\,P(XY)}{P(X)},\ \frac{P(XY)\,P(YZ)}{P(Y)},\ \frac{P(ZX)\,P(XY)}{P(Z)},$$

$$\frac{P(Y)\,P(Z)\,P(XYZ)}{P(YZ)},\ \frac{P(Z)\,P(X)\,P(XYZ)}{P(ZX)},\ \frac{P(X)\,P(Y)\,P(XYZ)}{P(XY)},$$

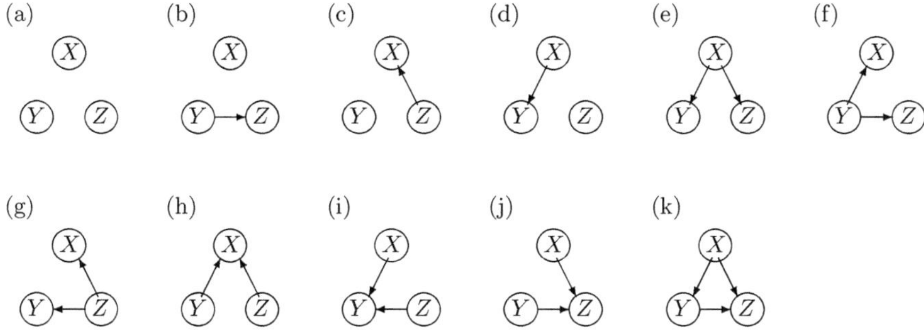

Fig. 3.6 The 11 Markov equivalent Bayesian networks for $p = 3$

$$P(XYZ).$$

Among these, in (a) one can simultaneously represent multiple conditional independencies such as

$$X \perp\!\!\!\perp Y, \quad Y \perp\!\!\!\perp Z, \quad Z \perp\!\!\!\perp X, \quad X \perp\!\!\!\perp \{Y, Z\}, \quad Y \perp\!\!\!\perp \{Z, X\}, \quad Z \perp\!\!\!\perp \{X, Y\},$$

while in (k), no nontrivial conditional independencies are represented. ■

For a given set of conditional independencies, sometimes a BN is suitable, and other times an MN is more appropriate.

Example 25 (When a BN is Appropriate) In Figs. 3.6 h, i, and j, respectively,

$$Y \perp\!\!\!\perp_G Z, \qquad Z \perp\!\!\!\perp_G X, \qquad X \perp\!\!\!\perp_G Y$$

hold, and from the factorized forms, we also have

$$Y \perp\!\!\!\perp Z, \qquad Z \perp\!\!\!\perp X, \qquad X \perp\!\!\!\perp Y.$$

Thus, (3.10) holds in each case. However, an MN cannot represent these nontrivial (here, marginal) independencies. Indeed, there are eight MNs on $p = 3$ (Fig. 3.7). In (h), no nontrivial conditional independence is asserted, so (3.9) holds trivially. Removing any edge from (h) yields one of (e), (f), or (g), which assert, respectively,

$$Y \perp\!\!\!\perp Z \mid X, \qquad Z \perp\!\!\!\perp X \mid Y, \qquad X \perp\!\!\!\perp Y \mid Z,$$

and these do not follow from $Y \perp\!\!\!\perp Z$, etc. Hence, no edges can be removed from (h). Therefore, the MN representing these independencies is (h), which asserts no nontrivial conditional independencies. ■

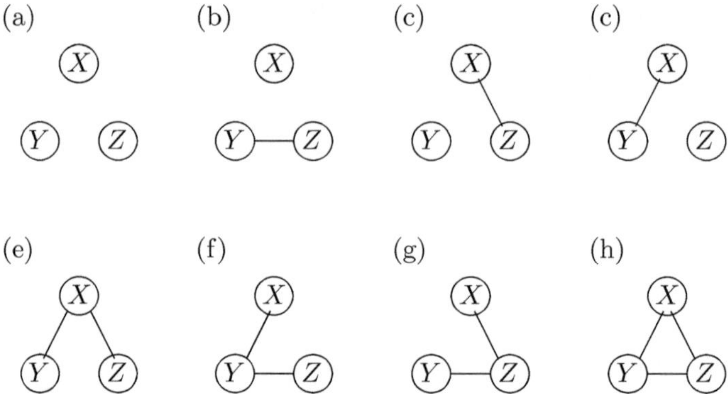

Fig. 3.7 The eight Markov networks for $p = 3$

Example 26 (When an MN is Appropriate) Consider random variables X, Y, Z, W with joint distribution

$$P(X, Y, Z, W) = \frac{P(X, Z)\, P(Z, Y)\, P(Y, W)\, P(W, X)}{P(X)\, P(Y)\, P(Z)\, P(W)}.$$

Then the conditional independencies $X \perp\!\!\!\perp Y \mid \{Z, W\}$ and $Z \perp\!\!\!\perp W \mid \{X, Y\}$ hold. Indeed,

$$P(X, Y, Z, W) = \frac{1}{P(Z)P(W)} \cdot \frac{P(W, X)\, P(X, Z)}{P(X)} \cdot \frac{P(W, Y)\, P(Y, Z)}{P(Y)},$$

and marginalizing over X, Y, and (X, Y) yields, respectively,

$$\begin{cases} P(Y, Z, W) = \dfrac{P(W, Y)\, P(Y, Z)}{P(Y)}, \\ P(X, Z, W) = \dfrac{P(W, X)\, P(X, Z)}{P(X)}, \end{cases} \qquad P(Z, W) = P(Z)\, P(W). \tag{3.11}$$

Therefore,

$$P(X, Y \mid Z, W) = P(X \mid Z, W)\, P(Y \mid Z, W) \tag{3.12}$$

holds. A similar derivation gives $Z \perp\!\!\!\perp W \mid \{X, Y\}$ (see Problem 26). In the MN of Fig. 3.8a), we have $X \perp\!\!\!\perp_G Y \mid \{Z, W\}$ and $Z \perp\!\!\!\perp_G W \mid \{X, Y\}$, so both (3.9) and (3.10) hold. However, attempting the same with a BN while avoiding cycles yields Fig. 3.8b, where only $Z \perp\!\!\!\perp_G W \mid \{X, Y\}$ holds, and Fig. 3.8c and d, where only $X \perp\!\!\!\perp_G Y \mid \{Z, W\}$ holds. Removing any further edges would assert independencies that do not hold. Hence, there is no BN for which both (3.9) and (3.10) hold simultaneously. ■

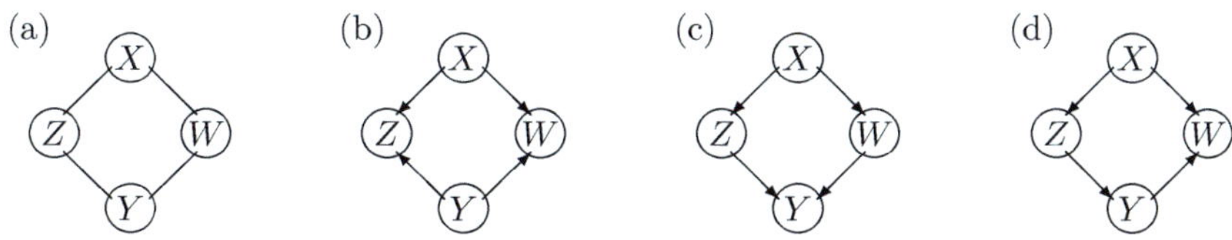

Fig. 3.8 In (**a**), both $X \perp\!\!\!\perp_G Y \mid \{Z, W\}$ and $Z \perp\!\!\!\perp_G W \mid \{X, Y\}$ hold; in (**c**), only $X \perp\!\!\!\perp_G Y \mid \{Z, W\}$ holds; in (**b**) and (**d**), only $Z \perp\!\!\!\perp_G W \mid \{X, Y\}$ holds

In later chapters, when both (3.9) and (3.10) hold for a BN, we say that the set of random variables is *faithful* to that BN, meaning that all and only the conditional independencies of the distribution are represented by the BN.

3.4 The Chow–Liu Algorithm

Consider an undirected graph G with vertex set $V = \{1, \dots, p\}$. For each unordered pair $i \neq j$, assign a weight $w_{i,j}$, and seek an edge set E maximizing the sum of weights. Assume the weights are positive and symmetric ($w_{i,j} = w_{j,i}$).

In particular, suppose $G = (V, E)$ is a *spanning tree*, i.e., a connected acyclic undirected graph. It is known that the following procedure [13] finds an optimal G (note that if multiple pairs share the same weight, the optimal tree need not be unique).

Kruskal Algorithm Initialize $\mathcal{E} = \{\{i, j\} \mid i \neq j\}$ and $E = \emptyset$. Repeat until $\mathcal{E} = \emptyset$:

1. Remove from $\mathcal{E}$ an edge $\{i, j\}$ with the largest weight $w_{i,j}$ (equivalently, process edges in descending order of weight).
2. If adding $\{i, j\}$ to E does not create a cycle, add it to E.

Example 27 For $V = \{1, \dots, 4\}$ with weights satisfying $w_{1,2} > w_{2,3} > w_{1,3} > w_{1,4} > w_{2,4} > w_{3,4} > 0$ (all other entries zero), the forest evolves in the order shown in Fig. 3.9. ■

Proposition 8 *Kruskal algorithm yields a spanning tree whose total edge weight is maximized.*

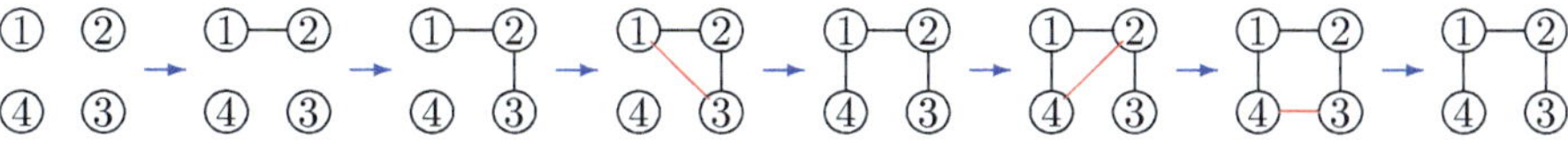

Fig. 3.9 Following the order of weights, edges {1, 2}, {2, 3}, {1, 3}, {1, 4}, {2, 4}, and {3, 4} are added in this sequence. However, when attempting to add {1, 3}, a cycle is created, so {1, 3} is not included. The same applies to {2, 4} and {3, 4}

Proof See [13] or the Wikipedia article "Kruskal Algorithm."

Kruskal algorithm can be adapted as follows without changing its behavior:

- Negative edge weights are allowed.
- One may obtain a maximum-weight forest instead of a spanning tree.

Since the goal is to maximize the total weight, vertices connected only by negative-weight edges will remain disconnected, yielding a forest rather than a tree.

In practice, *Prim algorithm* [21] is often preferred for efficiency.[1] See the appendix for details.

Next, for discrete random variables $X_1, \ldots, X_p$ with joint distribution

$$P(X_1 = x_1, \ldots, X_p = x_p), \qquad x_k \in \mathcal{X}_k,$$

where the sets of the values $X_1, \ldots, X_p$ take are denoted by $\mathcal{X}_1, \ldots, \mathcal{X}_p$, we consider approximating this distribution by one representable by a tree. Specifically, with $V = \{1, \ldots, p\}$ and $E \subset \{\{i, j\} \mid i \neq j\}$, approximate by

$$\prod_{k \in V} P(X_k = x_k) \prod_{\{i,j\} \in E} \frac{P(X_i = x_i, X_j = x_j)}{P(X_i = x_i)\, P(X_j = x_j)},$$

where E is chosen to be acyclic (a forest). For brevity, write $P(x_1, \ldots, x_p)$ for the original probability and $P'(x_1, \ldots, x_p)$ for the approximation. The Kullback–Leibler divergence between them is

$$D(P\|P') := \sum_{x_1} \cdots \sum_{x_p} P(x_1, \ldots, x_p) \log \frac{P(x_1, \ldots, x_p)}{P'(x_1, \ldots, x_p)}. \qquad (3.13)$$

Define the *mutual information* between X_i and X_j as

$$I(i, j) := \sum_{x_i} \sum_{x_j} P(x_i, x_j) \log \frac{P(x_i, x_j)}{P(x_i)\, P(x_j)}.$$

Let $H(k)$ denote the entropy of X_k, and $H(1, \ldots, p)$ denote the joint entropy. Then (see Problem 29),

$$D(P\|P') = -H(1, \ldots, p) + \sum_{k \in V} H(k) - \sum_{\{i,j\} \in E} I(i, j). \qquad (3.14)$$

The first two terms on the right-hand side of (3.14) do not depend on E, so minimizing $D(P\|P')$ amounts to maximizing the sum of mutual information in

[1] Historically, the algorithm was first published by Vojtěch Jarník in 1930; Prim re-published it in 1957.

the third term. Therefore, by taking weights $w_{i,j} = I(i, j)$ and applying Kruskal algorithm (or Prim Algorithm), we obtain the approximation P' minimizing the divergence. This procedure is the *Chow–Liu algorithm* [4].

Appendix

Prim Method

Given a subset $U \subset V$, at each step include in U a vertex $i \in V \setminus U$ maximizing $w[i, j]$ for some $j \in U$, and add the edge $\{i, j\}$.

Prim Algorithm Initialize $U = \emptyset$, $v[i] = 0$ $(i \in V)$, and $j \leftarrow 1$.

1. Add j to U. For each $i \notin U$ with $w[i, j] > v[i]$, set $v[i] \leftarrow w[i, j]$ and $\pi[i] \leftarrow j$.
2. Set j to be an $i \notin U$ maximizing $v[i]$.

Finally, the edges $\{i, \pi[i]\}$, $i = 2, \ldots, p$, form a maximum spanning tree.

Viewing a tree as rooted at some vertex, we may interpret edges as directed away from the root; the head i is the child and the tail $\pi[i]$ the parent. Any vertex can serve as the root (above, $j = 1$), and every non-root vertex has exactly one parent. Step 1 updates $v[i] \leftarrow \max_{j \in U} w[i, j]$ for each $i \notin U$ (and records $\pi[i]$ when updated). Prim method constructs, for each $1 \leq k \leq p$, a maximum-weight tree on a subset U of size k containing 1. This holds trivially for $|U| = 1$ $(U = \{1\})$. Assuming it holds for $|U| = k$, step 2 adds the edge with weight

$$\max_{i \notin U} \max_{j \in U} w[i, j] = \max_{i \notin U} w[i, \pi[i]] = \max_{i \notin U} v[i],$$

so the total weight remains maximal when $|U| = k + 1$ (each $i \notin U$ stores $\pi[i]$ attaining its current maximum). The claim also holds for $k = p$.

The above assumes a tree; for a forest, we look for isolated vertices and start a new set U.

The following is an R implementation (using the maximum total weight). Note it assumes a (possibly) disconnected graph and thus returns a maximum forest.

```
kruskal <- function(w) {      # Prim method (max-weight)
    p <- ncol(w)
    # Symmetrize
    for (i in 1:(p - 1)) {
        for (j in (i + 1):p) {
            w[j, i] <- w[i, j]
        }
    }
    # Zero diagonal
    for (i in 1:p) w[i, i] <- 0
```

```
    # Init
    parent <- array(dim = p)
    u <- rep(0, p)      # visited flags
    v <- rep(-Inf, p)   # best weights so far
    j <- 1
    parent[1] <- -1
    while (j <= p) {
        u[j] <- 1
        for (i in 1:p) {
            if (u[i] == 0 && w[j, i] > v[i]) {
                v[i] <- w[j, i]
                parent[i] <- j
            }
        }
        max_weight <- 0
        for (i in 1:p) {
            if (u[i] == 0 && v[i] > max_weight) {
                j <- i
                max_weight <- v[i]
            }
        }
        # If we cannot grow, start a new component
        if (max_weight <= 0) {
            j <- 1
            while (j <= p && u[j] == 1) j <- j + 1
            if (j <= p) parent[j] <- -1
        }
    }
    # Return edge list
    pair.1 <- NULL; pair.2 <- NULL
    for (j in 1:p) {
        if (parent[j] != -1) {
            pair.1 <- c(pair.1, parent[j])
            pair.2 <- c(pair.2, j)
        }
    }
    cbind(pair.1, pair.2)
}
```

In this code, multiple connected components (a forest) are allowed. If a spanning tree is guaranteed, the block `if(max_weight <= 0)` can be removed.

Example 28 Using `kruskal`, we compute the edge list (`edgelist`) and visualize it with the `igraph` package. Example outputs are shown in Fig. 3.10.

```
library(igraph)
p <- 10
w <- matrix(rnorm(p*p)^2, p, p)
edgelist <- kruskal(w)
plot(graph_from_edgelist(edgelist), directed=FALSE)
```

■

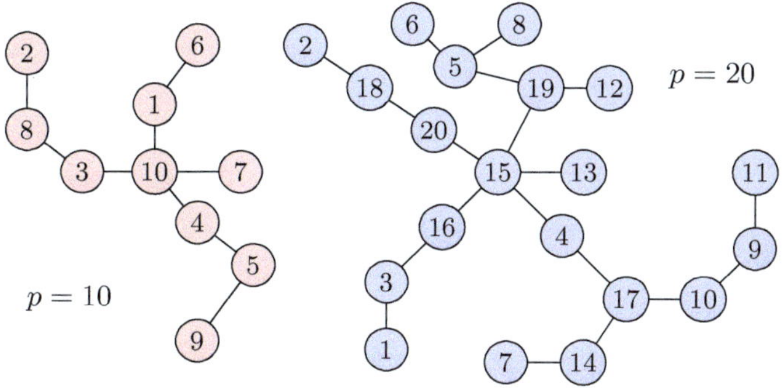

Fig. 3.10 Graphs obtained by running the program in Example 28. Left: $p = 10$; right: $p = 20$. Both are spanning trees

Problems 19–29

19. For each of the 11 Bayesian networks (BNs) in Fig. 3.6, identify which of the 25 directed acyclic graphs (DAGs) they represent.
20. The distributions of the 11 BNs in Fig. 3.6 are given in Example 24. What are the distributions of the eight Markov networks (MNs) in Fig. 3.7?
21. When a collider occurs at k on a path, state whether k d-separates in each of the following four cases:

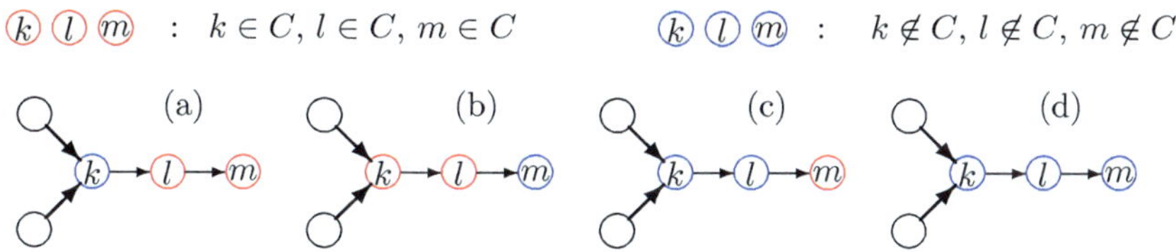

22. For the undirected graph in Fig. 3.5a, determine whether each of the following propositions is true or false:

    (a) $3 \perp\!\!\!\perp_G 9 \mid \{5, 8\}$
    (b) $7 \perp\!\!\!\perp_G 12 \mid \{3, 8\}$
    (c) $9 \perp\!\!\!\perp_G 12 \mid 8$
    (d) $\{1, 2\} \perp\!\!\!\perp_G 5 \mid 3$
    (e) $\{4, 5\} \perp\!\!\!\perp_G 6 \mid 7$
    (f) $\{7, 9\} \perp\!\!\!\perp_G 10 \mid 8$
    (g) $\{11, 9\} \perp\!\!\!\perp_G 12 \mid 13$
    (h) $\{1, 13\} \perp\!\!\!\perp_G 9 \mid 4$

23. For the DAG in Fig. 3.5b, determine whether each of the following propositions is true or false;

    (a) $3 \perp\!\!\!\perp_G 10 \mid \{5, 8\}$

(b) $7 \perp\!\!\!\perp_G 12 \mid \{3, 8\}$
(c) $5 \perp\!\!\!\perp_G 12 \mid 9$
(d) $1 \perp\!\!\!\perp_G 12 \mid \{4, 8\}$
(e) $2 \perp\!\!\!\perp_G 9 \mid \{5, 8\}$
(f) $3 \perp\!\!\!\perp_G 10 \mid 5$
(g) $7 \perp\!\!\!\perp_G 12 \mid \{3, 9\}$
(h) $11 \perp\!\!\!\perp_G 9 \mid 10$

24. Show that even if edges are added to Fig. 3.8b, c, and d in Example 26, the same separation relations as in Fig. 3.8a cannot be obtained. Further, show that if edges are deleted from (b)(c)(d), this would imply false conditional independencies.
25. Prove Proposition 7.
26. Explain why marginalization yields (3.11), and then use (3.11) to show (3.12).
27. Let X and Y be independent random variables taking values ± 1, and define $Z = XY$. Determine the Markov network (MN) and Bayesian network (BN) of X, Y, Z.
28. In the function `kruskal`:

    (a) If the input `w` is not symmetric, what does `w` become after lines 2 to 13?
    (b) What is the relationship between whether a vertex j belongs to the set U and the value of `u[j]`?
    (c) Between lines 39 and 48, when is -1 assigned to `parent[j]` in terms of the value of `u[j]`? Also, how is vertex `j` treated in that case?

29. (a) Show that (3.13) takes nonnegative values and equals 0 only when P and P' coincide. Also, state under what conditions (3.13) takes finite values.
    (b) Show that the mutual information $I(i, j)$ corresponds to a Kullback–Leibler divergence and hence is nonnegative. Also, state when the mutual information equals 0.
    (c) Show equality (3.14).

Chapter 4
Testing Independence and Conditional Independence with Kernels

In this chapter, we explain the methods for testing independence and conditional independence between variables using kernel functions. Kernel methods are powerful techniques that flexibly handle nonlinear and high-dimensional dependence structures.

We begin with an intuitive introduction to the basic concepts of reproducing kernel Hilbert spaces (RKHS). We then present the Hilbert–Schmidt independence criterion (HSIC) for testing the independence of two variables, along with its theoretical background and implementation. Next, we turn to Kernel Conditional Independence (KCI), which tests independence conditional on a third variable.

These methods construct test statistics by exploiting covariance structures in the feature spaces defined by kernels. We also discuss estimation procedures and implementation caveats for real-data applications.

The kernel-based tests introduced in this chapter play a central role in estimating causal structures in Chaps. 5 and 6.

4.1 Reproducing Kernel Hilbert Spaces

Let $\mathcal{X}$ be a set. A mapping $k : \mathcal{X} \times \mathcal{X} \to \mathbb{R}$ is called a *positive definite kernel* if it satisfies the following two conditions:

1. Symmetry: $k(x, y) = k(y, x)$ for $x, y \in \mathcal{X}$.
2. Positive definiteness: For any $N \geq 1$ and $x_1, \ldots, x_N \in \mathcal{X}$, the matrix $K \in \mathbb{R}^{N \times N}$ with entries $k(x_i, x_j)$ is positive semidefinite.

In particular, the matrix K above is called the *Gram matrix*. Note that positive definiteness here requires the Gram matrix to be positive semidefinite rather than strictly positive definite.

J. Suzuki, *Graphical Models and Causal Discovery with R*,
https://doi.org/10.1007/978-981-95-4267-3_4

In general, the term "kernel" is often used to refer to a measure of similarity within $\mathcal{X}$, and kernels that do not satisfy positive definiteness are also used in practice.

Example 29 (Epanechnikov Kernel) Let $\lambda > 0$. For $x, y \in \mathbb{R}$, define

$$k(x, y) = D\left(\frac{|x-y|}{\lambda}\right),$$

$$D(t) = \begin{cases} \frac{3}{4}(1-t^2), & |t| \le 1, \\ 0, & \text{otherwise.} \end{cases}$$

Apply this kernel to the smoothing method known as the *Nadaraya–Watson estimator*. Let observed data $(x_1, y_1), \ldots, (x_N, y_N) \in \mathcal{X} \times \mathbb{R}$ be given, and suppose a new $x_* \in \mathcal{X}$ is provided. Then return as $\hat{f}(x_*)$ the weighted average of $y_1, \ldots, y_N$ with weights

$$\frac{k(x_*, x_1)}{\sum_{j=1}^N k(x_*, x_j)}, \ldots, \frac{k(x_*, x_N)}{\sum_{j=1}^N k(x_*, x_j)}.$$

Since $k(x, y)$ is assumed to be large when $x, y \in \mathcal{X}$ are similar, larger weight is assigned to y_i whose x_i is more similar to x_*. As λ decreases, the prediction uses only (x_i, y_i) in a neighborhood of x_*. Figure 4.1 shows the result of this procedure. ■

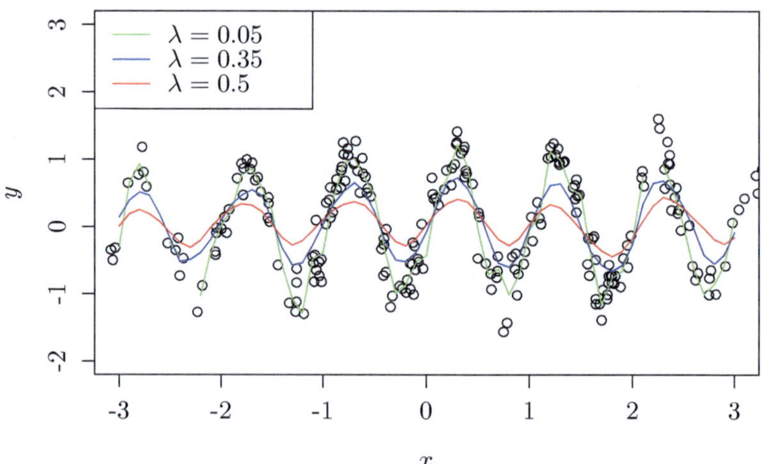

Fig. 4.1 Applying the Epanechnikov kernel to the Nadaraya–Watson estimator; curves are drawn for $\lambda = 0.05, 0.35, 0.5$

The kernel in Example 29 is not positive definite. For instance, with $\lambda = 2$, $n = 3$, and $x_1 = -1,\ x_2 = 0,\ x_3 = 1$,

$$K_\lambda = \begin{bmatrix} 3/4 & 9/16 & 0 \\ 9/16 & 3/4 & 9/16 \\ 0 & 9/16 & 3/4 \end{bmatrix}$$

one of the eigenvalues is negative (Problem 31).

On the other hand, the Gaussian kernel

$$k(x, y) = \exp\left\{-\frac{1}{2\sigma^2}\|x - y\|^2\right\}, \qquad \sigma^2 > 0,$$

and the Cauchy kernel

$$k(x, y) = \frac{1}{2\pi} \cdot \frac{1}{\|x - y\|^2 + \beta^2}, \qquad \beta > 0,$$

are positive definite when $\mathcal{X} = \mathbb{R}^p$. Here, for $a = [a_1, \ldots, a_p]$, set $\|a\| = \sqrt{\sum_{j=1}^p a_j^2}$. Positive definiteness follows from the fact that these $k(x, y)$ depend only on $\|x - y\|$ and are constant multiples of characteristic functions of probability densities,

$$f_X(x) = \frac{1}{\sqrt{2\pi\sigma^2}} \exp\left\{-\frac{x^2}{2\sigma^2}\right\}, \qquad f_X(x) = \frac{\beta}{2} \exp\{-\beta|x|\},$$

namely

$$\phi(t) = \int_{\mathcal{X}} e^{it^\top x} f_X(x)\, dx$$

(Problem 32). Indeed, the following proposition holds.

Proposition 9 (Bochner's Theorem [2]) *For any* $N \geq 1$, $x_1, \ldots, x_N \in \mathcal{X}$, *and* $z_1, \ldots, z_N \in \mathbb{R}$,

$$\sum_{i=1}^N \sum_{j=1}^N z_i z_j \phi(x_i - x_j) \geq 0$$

holds if and only if $\phi(\cdot)$ *is a constant multiple of the characteristic function of some probability density.*[1]

[1] A proof can be found in J. Suzuki, *100 Kernel Problems for Machine Learning with R/Python* [36, 37].

In addition, the inner product of a finite-dimensional Euclidean space also satisfies the definition of a kernel (Problem 33).

We will discuss later why positive definiteness is required for kernels; first, we record some properties.

For each $x \in \mathcal{X}$, consider the function

$$\mathcal{X} \ni y \mapsto k(x, y) \in \mathbb{R}.$$

Let H_0 be the set of functions f expressible as $f = \sum_{i=1}^{N} \alpha_i k(x_i, \cdot)$ for some $N \geq 1, x_1, \ldots, x_N \in \mathcal{X}$, and $\alpha_1, \ldots, \alpha_N \in \mathbb{R}$. Then H_0 is a *linear space*. That is, for any $\alpha \in \mathbb{R}$ and $f, g \in H_0$,

$$\alpha f \in H_0, \qquad f + g \in H_0 \tag{4.1}$$

(Problem 34). Define the *inner product* on H_0 by $\langle f, g \rangle = \sum_{i=1}^{N} \sum_{j=1}^{N} \alpha_i \beta_j k(x_i, x_j)$ for $f = \sum_{i=1}^{N} \alpha_i k(x_i, \cdot)$ and $g = \sum_{j=1}^{N} \beta_j k(x_j, \cdot)$. Then the four axioms of inner products hold for all $\alpha, \beta \in \mathbb{R}$ and $f, g, h \in H_0$:

$$\begin{aligned} &\langle f, f \rangle \geq 0, \quad \langle \alpha f + \beta g, h \rangle = \alpha \langle f, h \rangle + \beta \langle g, h \rangle, \quad \langle f, g \rangle = \langle g, f \rangle, \\ &\langle f, f \rangle = 0 \Longrightarrow f = 0. \end{aligned} \tag{4.2}$$

(Problem 34). In this book, we treat general linear spaces and inner products satisfying (4.1) and (4.2), rather than special cases such as the following example.

Example 30 (Two-Dimensional Euclidean Space) For $a_1, a_2, b_1, b_2 \in \mathbb{R}$ and $\alpha \in \mathbb{R}$,

$$\begin{bmatrix} a_1 \\ a_2 \end{bmatrix} + \begin{bmatrix} b_1 \\ b_2 \end{bmatrix} = \begin{bmatrix} a_1 + b_1 \\ a_2 + b_2 \end{bmatrix}, \qquad \alpha \begin{bmatrix} a_1 \\ a_2 \end{bmatrix} = \begin{bmatrix} \alpha a_1 \\ \alpha a_2 \end{bmatrix}.$$

This set of vectors forms a linear space, and the inner product of $\begin{bmatrix} a_1 \\ a_2 \end{bmatrix}$ and $\begin{bmatrix} b_1 \\ b_2 \end{bmatrix}$ is $a_1 b_1 + a_2 b_2$. ■

Next we define completion. The limit of a sequence of rational numbers $\{a_n\}$ need not be rational. However, even in such cases, for sufficiently large m, n, we have $|a_m - a_n| \to 0$.

Example 31 Consider $a_1 = 1$ and $a_{n+1} = \dfrac{1}{2} a_n + \dfrac{1}{a_n}$. Its limit is $\sqrt{2} \notin \mathbb{Q}$. Moreover,

$$|a_m - a_n| \leq |a_m - \sqrt{2}| + |a_n - \sqrt{2}| \to 0.$$

■

A sequence for which $|a_m - a_n| \to 0$ as m, n become large is called a **Cauchy sequence**. Without proof, any Cauchy sequence of real numbers is known to converge in $\mathbb{R}$.

In a set equipped with a distance (a *metric space*), if every Cauchy sequence converges within the set with respect to that distance, the space is called *complete*. In Example 31, the absolute value defines the distance on $\mathbb{R}$. The rationals $\mathbb{Q}$ are dense in $\mathbb{R}$, that is, for any $a \in \mathbb{R}$ and $\epsilon > 0$, the interval $(a - \epsilon, a + \epsilon)$ contains a rational number. It is known that $\mathbb{R}$ is obtained by adding points to $\mathbb{Q}$ to make it complete.

A linear space H equipped with an inner product is called an *inner-product space*. The norm induced by the inner product $\langle \cdot, \cdot \rangle_H$ is $\|a\|_H = \sqrt{\langle a, a \rangle_H}$ for $a \in H$.

Proposition 10 *Let H_1 be an inner-product space with the distance induced by its inner product. Then there exists a complete inner-product space H_2 containing H_1 as a dense subspace, and such an H_2 is unique up to isomorphism.*[2,3]

The operation that obtains H_2 from H_1 in Proposition 10 is called *completion*. A complete inner-product space is called a *Hilbert space*. The space H_2 consists of the limits of Cauchy sequences in H_1.

Example 32 (L^2 Space) Let L^2 denote the set of functions $f : [-\pi, \pi] \to \mathbb{R}$ satisfying

$$\int_{-\pi}^{\pi} \{f(x)\}^2 \, dx < \infty.$$

Then L^2 is a linear space (Problem 35), and with the inner product

$$\langle f, g \rangle = \int_{-\pi}^{\pi} f(x) g(x) \, dx,$$

it is complete. That is, for any sequence $\{f_n\} \subset L^2$ such that

$$\|f_m - f_n\| = \sqrt{\int_{-\pi}^{\pi} \{f_m(x) - f_n(x)\}^2 \, dx} \to 0 \quad (m, n \to \infty),$$

there exists $f \in L^2$ with $f_n \to f$. ∎

[2] Two inner-product spaces are isomorphic if there exists a bijective linear isometry between them.

[3] For a proof, see, e.g., https://bayesnet.org/books_jp/?p=1656.

Example 33 (Fourier Series Expansion) For a periodic function $f(x)$, we seek coefficients $a_0, a_1, \ldots$ and $b_1, b_2, \ldots$ such that

$$f(x) = a_0 + \sum_{n=1}^{\infty}(a_n \cos nx + b_n \sin nx).$$

For example, define

$$f(x) = \begin{cases} 1, & 0 \le x < \pi, \\ 0, & \pi \le x < 2\pi, \end{cases}$$

and extend periodically by $f(x + 2\pi) = f(x)$. Then

$$f(x) = \frac{1}{2} + \frac{2}{\pi}\Big(\sin x + \frac{1}{3}\sin 3x + \frac{1}{5}\sin 5x + \cdots\Big).$$

Let

$$f_N(x) = \frac{1}{2} + \frac{2}{\pi}\Big(\sin x + \frac{1}{3}\sin 3x + \cdots + \frac{1}{2N-1}\sin\{(2N-1)x\}\Big).$$

Figure 4.2 shows the graphs for $N = 1, 2, 3, 4, 5, 6$. For any finite N we do not obtain $f(x)$ exactly. Moreover, $f(x)$ cannot be expressed as a finite linear combination of $\cos nx$ and $\sin nx$ ($n = 1, 2, \ldots$). ■

For the inner-product space H_0 introduced in this section, there exists a Hilbert space H obtained by completing H_0. In particular, a Hilbert space obtained from a positive definite kernel is called a *reproducing kernel Hilbert space* (RKHS). The elements of H consist of limits of Cauchy sequences $\{f_n\}$ in H_0. For $f, g \in H$,

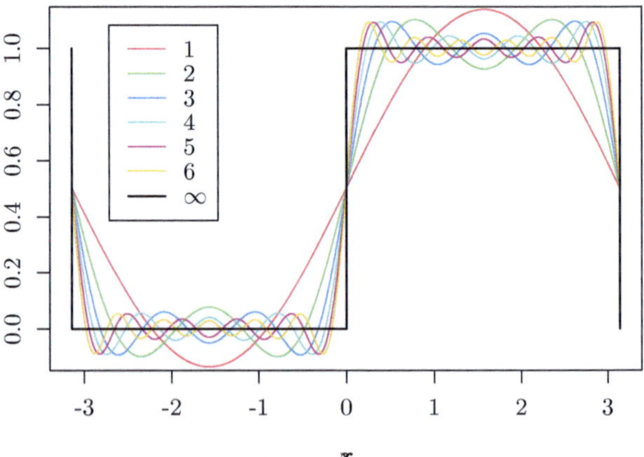

Fig. 4.2 Sketches of $f(x)$ ($N = \infty$) and $f_N(x)$ for $N = 1, \ldots, 6$

$$\langle f, g\rangle_H = \big\langle \lim_{N\to\infty} f_N,\ \lim_{N\to\infty} g_N\big\rangle_H := \lim_{N\to\infty} \langle f_N, g_N\rangle,$$

where $\langle\cdot,\cdot\rangle_H$ denotes the inner product in H. Even functions that are not finite linear combinations of $k(x,\cdot)$ can be represented as limits and belong to H. Linearity is preserved under taking limits, and the inner product is defined as the limit of inner products in H_0.

An RKHS H and a positive definite kernel k are in one-to-one correspondence, and the following identity—known as the *kernel trick*—holds:

$$\langle f, k(x,\cdot)\rangle_H = f(x), \qquad x \in \mathcal{X},\ f \in H. \tag{4.3}$$

Since $f \in H$ is given as $\lim_{N\to\infty} f_N$ for some Cauchy sequence $\{f_N\} \subset H_0$, we obtain (4.3) as follows: writing $f_N(\cdot) = \sum_{i=1}^N \alpha_i k(x_i,\cdot)$,

$$\begin{aligned}
\langle f, k(x,\cdot)\rangle_H &= \big\langle \lim_{N\to\infty} f_N,\ k(x,\cdot)\big\rangle_H = \lim_{N\to\infty} \langle f_N, k(x,\cdot)\rangle_{H_0} \\
&= \lim_{N\to\infty} \left\langle \sum_{i=1}^N \alpha_i k(x_i,\cdot),\ k(x,\cdot)\right\rangle_{H_0} = \lim_{N\to\infty} \sum_{i=1}^N \alpha_i k(x_i, x) \\
&= \lim_{N\to\infty} f_N(x) = f(x).
\end{aligned}$$

In data science and machine learning, kernels are used via the *feature map*

$$\mathcal{X} \ni x \mapsto k(x,\cdot) \in H,$$

which induces nonlinear representations. Concrete examples are given in Sects. 4.2 and 4.3 of this chapter.

4.2 Testing Independence: HSIC

Let $\mathcal{X}$ and $\mathcal{Y}$ be sets. Let $k_{\mathcal{X}} : \mathcal{X} \times \mathcal{X} \to \mathbb{R}$ and $k_{\mathcal{Y}} : \mathcal{Y} \times \mathcal{Y} \to \mathbb{R}$ be kernels, and let $H_{\mathcal{X}}$ and $H_{\mathcal{Y}}$ denote the corresponding RKHSs. In this section, we reduce the independence of $X \in \mathcal{X}$ and $Y \in \mathcal{Y}$ to the independence of $k_{\mathcal{X}}(X,\cdot) \in H_{\mathcal{X}}$ and $k_{\mathcal{Y}}(Y,\cdot) \in H_{\mathcal{Y}}$. Although $cov(X, Y) = 0$ does not imply independence of X and Y, here we define an analogue of covariance for $k_{\mathcal{X}}(X,\cdot)$ and $k_{\mathcal{Y}}(Y,\cdot)$, and test whether it is zero.

In general, the product of two kernels is again positive definite. This follows from the proposition below.

Proposition 11 (Schur Product Theorem) *If $A = (a_{ij})$ and $B = (b_{ij})$ are positive semidefinite, then the Hadamard product $C = (a_{ij}b_{ij})$ is also positive semidefinite.*

Thus, the RKHS $H_{\mathcal{X}\mathcal{Y}}$ corresponding to the positive definite kernel $k_{\mathcal{X}}k_{\mathcal{Y}}$ exists. By definition,

$$k_{\mathcal{X}\mathcal{Y}}(x, y, \cdot, \star) = k_{\mathcal{X}}(x, \cdot)\, k_{\mathcal{Y}}(y, \star), \qquad x \in \mathcal{X},\ y \in \mathcal{Y},$$

is an element of $H_{\mathcal{X}\mathcal{Y}}$.[4] Define the means

$$m_X := \mathbb{E}_X[k_{\mathcal{X}}(X, \cdot)], \quad m_Y := \mathbb{E}_Y[k_{\mathcal{Y}}(Y, \star)], \quad m_{XY} := \mathbb{E}_{XY}[k_{\mathcal{X}\mathcal{Y}}(X, Y, \cdot, \star)],$$

which belong to $H_{\mathcal{X}}$, $H_{\mathcal{Y}}$, and $H_{\mathcal{X}\mathcal{Y}}$, respectively. Since products of elements in $H_{\mathcal{X}}$ and $H_{\mathcal{Y}}$ lie in $H_{\mathcal{X}\mathcal{Y}}$, we may regard $m_X m_Y \in H_{\mathcal{X}\mathcal{Y}}$. In particular, we call $m_{XY} - m_X m_Y \in H_{\mathcal{X}\mathcal{Y}}$ the *covariance* of X and Y in the RKHS sense. Since $H_{\mathcal{X}\mathcal{Y}}$ is an RKHS, we can define

$$HSIC(X, Y) = \|m_{XY} - m_X m_Y\| \tag{4.4}$$

(the *Hilbert–Schmidt independence criterion*, HSIC). Gretton et al. [8] proposed using an estimator of this quantity to test independence. In fact:

Proposition 12 *If $k_{\mathcal{X}}$ and $k_{\mathcal{Y}}$ are both characteristic kernels, then*

$$X \perp\!\!\!\perp Y \iff HSIC(X, Y) = 0.$$

A *characteristic kernel* is one for which the mapping

$$\mathcal{P} \ni P \mapsto m_X = \int_{\mathcal{X}} k(x, \cdot)\, dP(x)$$

is injective on a family $\mathcal{P}$ of probability distributions for X (if a density exists, write $dP(x) = f_X(x)dx$). In particular, when $k(x, y)$ is a function of $\|x - y\|$, the following is known.

Proposition 13 *A kernel $k : \mathcal{X} \times \mathcal{X} \to \mathbb{R}$ is characteristic if and only if the support of the corresponding probability density (whose characteristic function is k) is $\mathcal{X}$.*

Here, the support of f_X is the closure of $\{x \in \mathcal{X} : f_X(x) \neq 0\}$. Gaussian and Laplace kernels correspond to the characteristic functions of the normal and Laplace distributions, whose supports are all of $\mathcal{X} = \mathbb{R}$, hence they are characteristic.

Since the joint distribution of (X, Y) is unknown, $HSIC(X, Y)$ must be estimated from observations $x_1, \ldots, x_n \in \mathcal{X}$ and $y_1, \ldots, y_n \in \mathcal{Y}$. For a statistical test, we need a statistic and its null distribution. A more explicit expansion of $HSIC(X, Y)$ is as follows.

[4] For functions of two variables, we use $\cdot$ and $\star$ to indicate the corresponding slots.

Proposition 14 *$HSIC(X, Y)$ expands as*

$$\begin{aligned}\|m_{XY} - m_X m_Y\|^2 = & \mathbb{E}_{XX'YY'}[k_{\mathcal{X}}(X, X')\, k_{\mathcal{Y}}(Y, Y')] \\ & - 2\,\mathbb{E}_{XY}\left\{\mathbb{E}_{X'}[k_{\mathcal{X}}(X, X')]\; \mathbb{E}_{Y'}[k_{\mathcal{Y}}(Y, Y')]\right\} \\ & + \mathbb{E}_{XX'}[k_{\mathcal{X}}(X, X')]\; \mathbb{E}_{YY'}[k_{\mathcal{Y}}(Y, Y')],\end{aligned}$$

where X and X', and Y and Y' are independent and identically distributed.

Proof See the appendix at the end of this chapter.

When applying HSIC, we typically replace expectations by empirical averages to obtain the estimator

$$\begin{aligned}\widehat{HSIC} := & \frac{1}{N^2}\sum_i\sum_j k_{\mathcal{X}}(x_i, x_j)\, k_{\mathcal{Y}}(y_i, y_j) \\ & -\frac{2}{N^3}\sum_i\left\{\sum_j k_{\mathcal{X}}(x_i, x_j)\sum_h k_{\mathcal{Y}}(y_i, y_h)\right\} \\ & +\frac{1}{N^4}\sum_i\sum_j k_{\mathcal{X}}(x_i, x_j)\sum_h\sum_r k_{\mathcal{Y}}(y_h, y_r),\end{aligned} \tag{4.5}$$

where all summations run from 1 to n.

For $x_1, \ldots, x_n \in \mathcal{X}$, define the *centralized Gram matrix* $\tilde{K} = (\tilde{k}_{\mathcal{X}}(x_i, x_j))$ by

$$\tilde{k}_{\mathcal{X}}(x_i, x_j) := \left\langle k_{\mathcal{X}}(x_i, \cdot) - \frac{1}{n}\sum_{s=1}^{n} k_{\mathcal{X}}(x_s, \cdot),\; k_{\mathcal{X}}(x_j, \cdot) - \frac{1}{n}\sum_{t=1}^{n} k_{\mathcal{X}}(x_t, \cdot)\right\rangle.$$

Let E be the $n \times n$ all-ones matrix and I the $n \times n$ identity; set $H := I - \frac{1}{n}E$. Then

$$\tilde{K}_{\mathcal{X}} = H K_{\mathcal{X}} H. \tag{4.6}$$

Indeed, with $\delta_{ij} = 1$ if $i = j$ and 0 otherwise,

$$\begin{aligned}& \sum_{s=1}^{n}\sum_{t=1}^{n}(\delta_{is} - \tfrac{1}{n})\, k_{\mathcal{X}}(x_s, x_t)\, (\delta_{tj} - \tfrac{1}{n}) \\ = & \sum_{s=1}^{n}(\delta_{is} - \tfrac{1}{n})\left\{k_{\mathcal{X}}(x_s, x_j) - \frac{1}{n}\sum_{t=1}^{n} k_{\mathcal{X}}(x_s, x_t)\right\} \\ = & k_{\mathcal{X}}(x_i, x_j) - \frac{1}{n}\sum_{t=1}^{n} k_{\mathcal{X}}(x_i, x_t) - \frac{1}{n}\sum_{s=1}^{n} k_{\mathcal{X}}(x_s, x_j) + \frac{1}{n^2}\sum_{s=1}^{n}\sum_{t=1}^{n} k_{\mathcal{X}}(x_s, x_t)\end{aligned}$$

$$= \left\langle k_{\mathcal{X}}(x_i, \cdot) - \frac{1}{n}\sum_{s=1}^{n} k_{\mathcal{X}}(x_s, \cdot),\ k_{\mathcal{X}}(x_j, \cdot) - \frac{1}{n}\sum_{t=1}^{n} k_{\mathcal{X}}(x_t, \cdot) \right\rangle,$$

where we used $\langle k_{\mathcal{X}}(x_i, \cdot), k_{\mathcal{X}}(x_j, \cdot)\rangle = k_{\mathcal{X}}(x_i, x_j)$. Hence (4.6) holds.

Moreover:

Proposition 15 *$\widehat{HSIC}$ can be written as*

$$\widehat{HSIC} = \frac{1}{n^2}\operatorname{tr}(\tilde{K}_{\mathcal{X}}\tilde{K}_{\mathcal{Y}}). \tag{4.7}$$

Proof See the appendix at the end of this chapter.

In R, the following functions implement these ideas:

```r
# Construct a kernel matrix
K <- function(k, X) {
  n <- if (is.vector(X)) length(X) else nrow(X)
  K <- matrix(0, n, n)
  for (i in 1:n) {
    for (j in 1:n) {
      K[i, j] <- if (is.vector(X)) k(X[i], X[j]) else k(X[i, ], X[j, ])
    }
  }
  return(K)
}
```

```r
# Center a kernel matrix
tilde <- function(K) {
  # K: n x n kernel matrix to be centered
  n <- nrow(K)
  H <- diag(n) - matrix(1/n, n, n)  # centering matrix
  return(H %*% K %*% H)             # centered kernel matrix
}
```

```r
# Compute HSIC (Hilbert--Schmidt Independence Criterion)
HSIC.1 <- function(K.x, K.y) {

  # K.x: kernel matrix of the first variable
  # K.y: kernel matrix of the second variable

  n <- ncol(K.x)  # sample size

  # Centered kernel matrices
  K.x_tilde <- tilde(K.x)
  K.y_tilde <- tilde(K.y)

  # HSIC value
  hsic_value <- sum(diag(K.x_tilde %*% K.y_tilde)) / n^2

  return(hsic_value)
}
```

For testing independence, the null hypothesis is that X and Y are independent. For the HSIC estimator, a statistic with a known exact null distribution is not available. One may use a *permutation test* or the asymptotic null distribution for large n.

In a permutation test for independence, although $(X, Y) = (x_1, y_1), \ldots, (x_n, y_n)$ may not be independent pairwise, shifted pairs such as $(x_1, y_2), \ldots, (x_{n-1}, y_n)$, (x_n, y_1) are independent. Generate many $\widehat{HSIC}$ values for such independent pairings to form the null distribution. If the observed $\widehat{HSIC}$ lies in the upper α tail, reject the null of independence (one-sided test).

Example 34 Using the Gaussian kernel with $\sigma^2 = 1$, we generated pairs of independent standard normal random variables $(x_1, y_1), \ldots, (x_n, y_n)$ ($n = 50$; Fig. 4.3a) and pairs $(x_1, x_1+y_1), \ldots, (x_n, x_n+y_n)$ (Fig. 4.3b). We estimated the null distribution and checked whether the observed $\widehat{HSIC}$ falls in the rejection region. With significance level $\alpha = 0.05$ and $m = 100$ permutations, we did not reject the null for the independent case but did reject for the dependent case. The R code used is shown below. ■

```
## Visualize HSIC's null distribution with a histogram ##

# Data generation
n <- 50
x <- rnorm(n)  # standard normal x
y <- rnorm(n)  # standard normal y

# Kernel (RBF/Gaussian) with variance parameter
sigma2 <- 1
k.x <- function(x, y) exp(-sum((x - y)^2) / (2 * sigma2))
k.y <- k.x

# Kernel matrices
K.x <- K(k.x, x)
K.y <- K(k.y, y)

# Observed HSIC
u <- HSIC.1(K.x, K.y)

# Permutation-based null distribution
m <- 100
w <- NULL
for (i in 1:m) {
  x_shuffled <- sample(x, n)
  K.x <- K(k.x, x_shuffled)
  w <- c(w, HSIC.1(K.x, K.y))
}

# Rejection threshold (alpha = 0.05)
v <- quantile(w, 0.95)

# Plot
```

```
plot(
  density(w),
  xlim = c(min(w, v, u), max(w, v, u)),
  main = "Null distribution of HSIC and the observed value",
  xlab = "HSIC",
  ylab = "Density"
)
abline(v = v, col = "red", lty = 2, lwd = 2)   # critical value
abline(v = u, col = "blue", lty = 1, lwd = 2)  # observed HSIC
```

Several approaches use asymptotic null distributions. Although Gretton et al. propose a U-statistic approach, here we introduce the method of Kun Zhang [43], which will connect to the next section.

Proposition 16 *Let* $\lambda_{\mathcal{X},1} \geq \cdots \geq \lambda_{\mathcal{X},n}$ *and* $\lambda_{\mathcal{Y},1} \geq \cdots \geq \lambda_{\mathcal{Y},n}$ *be the eigenvalues of* $\tilde{K}_{\mathcal{X}}$ *and* $\tilde{K}_{\mathcal{Y}}$*, respectively. Let* Z_{ij}*,* $i, j = 1, \ldots, n$*, be i.i.d.* $N(0, 1)$*. Under* $X \perp\!\!\!\perp Y$*, the statistic* $n\,\widehat{HSIC}$ *is asymptotically distributed as*

$$\frac{1}{n^2}\sum_{i=1}^{n}\sum_{j=1}^{n}\lambda_{\mathcal{X},i}\,\lambda_{\mathcal{Y},j}\,Z_{ij}^2.$$

Proof See the appendix at the end of this chapter.

Leaving the theory and proofs to the next section, we now conduct numerical experiments using this method. We extend the function HSIC.1 to also compute eigenvalues and return n^3 times the statistic (i.e., n times the trace) together with the eigenvalues.

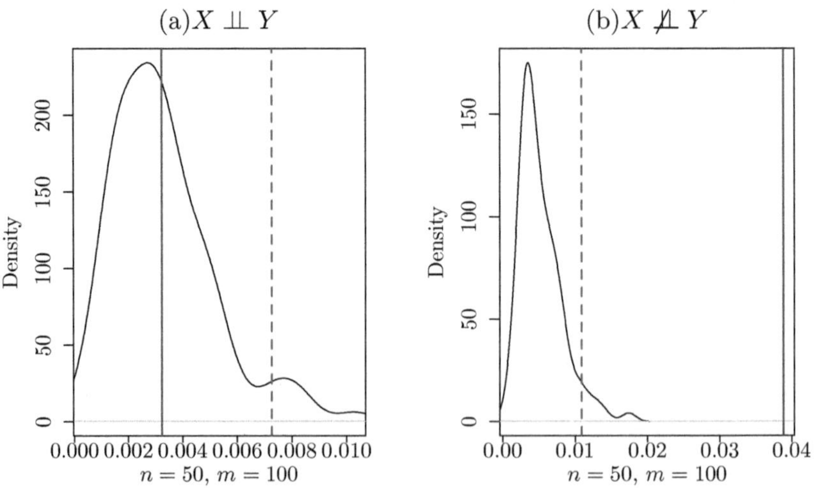

Fig. 4.3 Null distributions (curves) and observed $\widehat{HSIC}$ (blue vertical line) in Example 34. The red dashed line marks the rejection boundary. (**a**) Independent X, Y; (**b**) dependent X, Y

```
# Return n^3 * HSIC and eigenvalues
HSIC.2 <- function(K.x, K.y) {

  n <- ncol(K.x)

  # Centered kernel matrices
  tilde.x <- tilde(K.x)
  tilde.y <- tilde(K.y)

  # n^2 * HSIC (trace form)
  hsic_value <- sum(diag(tilde.x %*% tilde.y))

  # Eigenvalues
  lambda.x <- Re(eigen(tilde.x)$values)
  lambda.y <- Re(eigen(tilde.y)$values)
  lambda.xy <- sort(as.vector(outer(lambda.x, lambda.y, "*")), decreasing
    = TRUE)

  # Return n^3 * HSIC and eigenvalues
  return(list(statistics = hsic_value * n,
              eigen_values = lambda.xy))
}
```

Example 35 We computed the eigenvalues of $\tilde{K}_{\mathcal{X}}$ and $\tilde{K}_{\mathcal{Y}}$ and, following Proposition 16, generated 10,000 random values from the distribution of $n\,\widehat{HSIC}$ to obtain the upper $\alpha = 0.05$ critical value. As in Example 34, we generated data for the independent and dependent cases. The null was not rejected in the independent case and was rejected in the dependent case. The function `null.dist` below returns the simulated values z of $\sum_{i=1}^{n}\sum_{j=1}^{n}\lambda_{\mathcal{X},i}\lambda_{\mathcal{Y},j}Z_{ij}^2$ and its $100(1-\alpha)$-th percentile. ■

```
# Simulate null distribution and return critical value
null.dist <- function(eigen, alpha) {
  if (length(eigen) == 0) return(list(z = numeric(0), critical = NA))
  r <- length(eigen)

  z <- numeric(10000)
  for (h in 1:10000) {
    z[h] <- sum(eigen * rchisq(r, df = 1))
  }

  critical <- quantile(z, 1 - alpha)
  return(list(z = z, critical = critical))
}
```

Verification code (note that `null.dist` outputs the unscaled $\sum\lambda_{\mathcal{X},i}\lambda_{\mathcal{Y},j}Z_{ij}^2$):

```
# Sample size and data
n <- 50
x <- rnorm(n)
y <- rnorm(n)
# y <- x + rnorm(n)  # dependent case
```

```

# Kernel matrices
K.x <- K(k.x, x)
K.y <- K(k.y, y)

# HSIC and eigenvalues
result <- HSIC.2(K.x, K.y)
actual_value <- result$statistics
eigen <- result$eigen_values

# 95% critical value of the null distribution
critical <- null.dist(eigen, 0.05)$critical

cat("95% critical point of the null:", critical, "\n")
cat("Observed n * HSIC:", actual_value, "\n")
```

Example output for independent standard normals X, Y:

```
> 95% critical point of the null: 621.8676
> Observed n * HSIC: 253.9742
```

When setting `y <- rnorm(n) + x`:

```
> 95% critical point of the null: 743.1694
> Observed n * HSIC: 4964.367
```

4.3 Testing Conditional Independence: KCI

Let $\mathcal{X}$, $\mathcal{Y}$, and $\mathcal{Z}$ be sets, and suppose we have observations $(X, Y, Z) = (x_1, y_1, z_1), \ldots, (x_n, y_n, z_n) \in \mathcal{X} \times \mathcal{Y} \times \mathcal{Z}$. We test conditional independence $X \perp\!\!\!\perp Y \mid Z$. In this section, we introduce the *Kernel-based conditional independence (KCI)* test proposed by Kun Zhang et al. in 2011 [43].

Let $K_{\mathcal{Z}} \in \mathbb{R}^{n\times n}$ be the Gram matrix of a kernel $k_{\mathcal{Z}} : \mathcal{Z} \times \mathcal{Z} \to \mathbb{R}$. Let $k_{XZ} : \mathcal{X} \times \mathcal{Z} \times \mathcal{X} \times \mathcal{Z} \to \mathbb{R}$ be the product kernel of k_X and k_Z, and let $K_{\mathcal{X}\mathcal{Z}} \in \mathbb{R}^{n\times n}$ be its Gram matrix. Define the centered matrices $\tilde{K}_{\mathcal{Z}} := HK_{\mathcal{Z}}H$ and $\tilde{K}_{\mathcal{X}\mathcal{Z}} := HK_{\mathcal{X}\mathcal{Z}}H$. Define $\tilde{K}_{\mathcal{Y}\mathcal{Z}}$ analogously.

To construct a test statistic T for conditional independence, we introduce *kernel ridge regression*. Given data $x_1, \ldots, x_n \in \mathcal{X} = \mathbb{R}$ and $y_1, \ldots, y_n \in \mathbb{R}$, for $\lambda > 0$ consider

$$\sum_{i=1}^{n}\{y_i - f(x_i)\}^2 + \lambda\|f\|_H^2, \tag{4.8}$$

where H is an RKHS with norm $\|\cdot\|_H$ and kernel k. If $f \in H$, the solution is of the form $f(x) = \sum_{j=1}^{n}\alpha_j k(x, x_j)$ for some $\alpha = [\alpha_1, \ldots, \alpha_n]^\top$, with $\hat{\alpha} = (K + \lambda I)^{-1}[y_1, \ldots, y_n]^\top$ (Problem 38).

In matrix form,

$$\begin{bmatrix} \hat{f}(x_1) \\ \vdots \\ \hat{f}(x_n) \end{bmatrix} = \begin{bmatrix} \sum_{j=1}^n \hat{\alpha}_j k(x_1, x_j) \\ \vdots \\ \sum_{j=1}^n \hat{\alpha}_j k(x_n, x_j) \end{bmatrix} = K\hat{\alpha} = K(K + \lambda I)^{-1} \begin{bmatrix} y_1 \\ \vdots \\ y_n \end{bmatrix}.$$

The penalty term proportional to $\| f \|_H$ in (4.8) prevents overfitting by discouraging overly complex functions. With $\lambda = 0$ (ordinary least squares), the prediction $f(X)$ may be overly sensitive to X. Choosing $\lambda > 0$ appropriately yields predictions closer to the average response given X. Note how the $(K + \lambda I)^{-1}$ factor shrinks coefficients compared to K^{-1}.

We now apply this idea not to $\mathcal{X}$ and $\mathcal{Y}$ directly, but to

$$(\mathcal{X} \times \mathcal{Z}) \ni (x, z) \mapsto k_{\mathcal{X}\mathcal{Z}}((x, z), (\cdot, \star)) \in H_{\mathcal{X}\mathcal{Z}},$$
$$(\mathcal{Y} \times \mathcal{Z}) \ni (y, z) \mapsto k_{\mathcal{Y}\mathcal{Z}}((y, z), (\cdot, \star)) \in H_{\mathcal{Y}\mathcal{Z}},$$

and test conditional independence in $H_{\mathcal{X}\mathcal{Z}}$ and $H_{\mathcal{Y}\mathcal{Z}}$.[5] KCI uses the centered kernels $\tilde{k}_{\mathcal{X}\mathcal{Z}}$ and $\tilde{k}_{\mathcal{Y}\mathcal{Z}}$. For $f \in H_{\mathcal{X}\mathcal{Z}}$ and $g \in H_{\mathcal{Y}\mathcal{Z}}$, subtract the conditional mean given $Z = z$:

$$f(x, z) - \tilde{K}_{\mathcal{Z}}(\tilde{K}_{\mathcal{Z}} + \lambda I)^{-1} f(x, z) = R_{\mathcal{Z}}\, f(x, z),$$
$$g(y, z) - \tilde{K}_{\mathcal{Z}}(\tilde{K}_{\mathcal{Z}} + \lambda I)^{-1} g(y, z) = R_{\mathcal{Z}}\, g(y, z),$$

where

$$R_{\mathcal{Z}} := I - \tilde{K}_{\mathcal{Z}}(\tilde{K}_{\mathcal{Z}} + \lambda I)^{-1} = \lambda(\tilde{K}_{\mathcal{Z}} + \lambda I)^{-1}.$$

Thus, the feature maps

$$(x, z) \mapsto \begin{bmatrix} \tilde{k}_{\mathcal{X}\mathcal{Z}}((x_1, z_1), (x, z)) \\ \vdots \\ \tilde{k}_{\mathcal{X}\mathcal{Z}}((x_n, z_n), (x, z)) \end{bmatrix}, \qquad (y, z) \mapsto \begin{bmatrix} \tilde{k}_{\mathcal{Y}\mathcal{Z}}((y_1, z_1), (y, z)) \\ \vdots \\ \tilde{k}_{\mathcal{Y}\mathcal{Z}}((y_n, z_n), (y, z)) \end{bmatrix}$$

are replaced by

$$(x, z) \mapsto R_{\mathcal{Z}} \begin{bmatrix} \tilde{k}_{\mathcal{X}\mathcal{Z}}((x_1, z_1), (x, z)) \\ \vdots \\ \tilde{k}_{\mathcal{X}\mathcal{Z}}((x_n, z_n), (x, z)) \end{bmatrix}, \qquad (y, z) \mapsto R_{\mathcal{Z}} \begin{bmatrix} \tilde{k}_{\mathcal{Y}\mathcal{Z}}((y_1, z_1), (y, z)) \\ \vdots \\ \tilde{k}_{\mathcal{Y}\mathcal{Z}}((y_n, z_n), (y, z)) \end{bmatrix}.$$

[5] If $k : \mathcal{X} \times \mathcal{X} \to \mathbb{R}$ and $x \in \mathcal{X}$, then $k(x, \cdot)$ denotes the function $y \mapsto k(x, y)$. For $k : (\mathcal{X} \times \mathcal{Y}) \times (\mathcal{X} \times \mathcal{Y}) \to \mathbb{R}$ and $(x, y) \in \mathcal{X} \times \mathcal{Y}$, we write $k(x, y, \cdot, \star)$ for the function $(u, v) \mapsto k(x, y, u, v)$.

Hence, with $\tilde{K}_{\mathcal{XZ}} = V_{\mathcal{XZ}} V_{\mathcal{XZ}}^\top$ and $\tilde{K}_{\mathcal{YZ}} = V_{\mathcal{YZ}} V_{\mathcal{YZ}}^\top$,

$$\tilde{K}_{\mathcal{XZ}|\mathcal{Z}} = R_{\mathcal{Z}} V_{\mathcal{XZ}} \{R_{\mathcal{Z}} V_{\mathcal{XZ}}\}^\top = R_{\mathcal{Z}}\, \tilde{K}_{\mathcal{XZ}}\, R_{\mathcal{Z}},$$

$$\tilde{K}_{\mathcal{YZ}|\mathcal{Z}} = R_{\mathcal{Z}} V_{\mathcal{YZ}} \{R_{\mathcal{Z}} V_{\mathcal{YZ}}\}^\top = R_{\mathcal{Z}}\, \tilde{K}_{\mathcal{YZ}}\, R_{\mathcal{Z}}.$$

Analogously to the independence test, define the statistic

$$T := \frac{1}{n} \operatorname{tr}\Big(\tilde{K}_{\mathcal{XZ}|\mathcal{Z}}\, \tilde{K}_{\mathcal{YZ}|\mathcal{Z}}\Big),$$

and derive its asymptotic null distribution under $X \perp\!\!\!\perp Y \mid Z$.[6]

Let $R_{\mathcal{Z}} V_{\mathcal{XZ}} = \phi = (\phi_{ij})$ and $R_{\mathcal{Z}} V_{\mathcal{YZ}} = \varphi = (\varphi_{ik})$, so that $\tilde{K}_{\mathcal{XZ}|\mathcal{Z}} = \phi\phi^\top$ and $\tilde{K}_{\mathcal{YZ}|\mathcal{Z}} = \varphi\varphi^\top$. Define $w_i = (w_{i,j,k}) \in \mathbb{R}^{n^2}$ with entries $w_{i,j,k} = \phi_{ij}\varphi_{ik}$ and form the matrix $w = [w_1, \ldots, w_n] \in \mathbb{R}^{n^2 \times n}$. Then each entry of $ww^\top \in \mathbb{R}^{n^2 \times n^2}$ at index $((j,k),(j',k'))$ is

$$\sum_{i=1}^{n} \phi_{ij}\varphi_{ik}\phi_{ij'}\varphi_{ik'}.$$

Let $\lambda_1, \ldots, \lambda_{n^2}$ be the eigenvalues of $ww^\top$. Then:

Proposition 17 (Kun Zhang 2011) *Let $Z_1, \ldots, Z_{n^2} \sim N(0,1)$ be independent. Under $X \perp\!\!\!\perp Y \mid Z$ and as $n \to \infty$, the statistic T is distributed as*

$$\frac{1}{n} \sum_{k=1}^{n^2} \lambda_k Z_k^2.$$

Proof See the appendix at the end of this chapter.

An R implementation is as follows: We first force symmetry to avoid numerical issues (e.g., Cholesky failures) when matrices are not exactly symmetric.

```
symmetrize <- function(A){
  return( (t(A) + A) / 2 )
}
```

```
# KCI (Kernel-based conditional independence test)
KCI.1 <- function(K.x, K.y, K.z, lambda) {
  n <- nrow(K.x)
```

[6] The original KCI paper uses $T := \frac{1}{n}\operatorname{tr}(\tilde{K}_{\mathcal{XZ}|\mathcal{Z}} \tilde{K}_{\mathcal{Y}|\mathcal{Z}})$ and derives the null with $\mathcal{YZ}$ replaced by $\mathcal{Y}$. This is equivalent to the treatment here. The original form is computationally convenient; we adopt a symmetric form for clarity.

```
4
5    # Joint kernels with z
6    K.xz <- K.x * K.z
7    K.yz <- K.y * K.z
8
9    # Regularized inverse related term for z
10   R.Z <- lambda * solve(tilde(K.z) + lambda * diag(n))
11
12   # Project out z
13   XZ <- R.Z %*% tilde(K.xz) %*% t(R.Z)
14   YZ <- R.Z %*% tilde(K.yz) %*% t(R.Z)
15
16   # Eigen-decompositions
17   eigen.XZ <- eigen(symmetrize(XZ))
18   eigen.YZ <- eigen(symmetrize(YZ))
19
20   # Use top r eigen-components
21   r <- ceiling(sqrt(n))
22   S <- tcrossprod(eigen.XZ$vectors[, 1:r], diag(sqrt(Re(eigen.XZ$values[1:r]))))
23   T <- tcrossprod(eigen.YZ$vectors[, 1:r], diag(sqrt(Re(eigen.YZ$values[1:r]))))
24
25   # Build W (n x r^2) with column-wise products
26   W <- matrix(0, ncol = r*r, nrow = n)
27   for (i in 1:r) {
28     for (j in 1:r) {
29       W[, r * (i - 1) + j] <- S[, i] * T[, j]
30     }
31   }
32   return(list(
33     statistics = sum(diag(XZ %*% YZ)),         # test statistic T
34     eigen_values = Re(eigen(crossprod(W))$values)  # eigenvalues of W^T W
35   ))
36 }
```

Up to line 14 we have

$$\tilde{K}_{\mathcal{X}\mathcal{Z}|\mathcal{Z}} = \phi\phi^\top, \qquad \tilde{K}_{\mathcal{Y}\mathcal{Z}|\mathcal{Z}} = \varphi\varphi^\top.$$

In lines 17–18 we write

$$\phi = U\Lambda_X^{1/2}, \quad U = [u_1, \dots, u_n] = \begin{bmatrix} u_{11} & \cdots & u_{1n} \\ \vdots & \ddots & \vdots \\ u_{n1} & \cdots & u_{nn} \end{bmatrix}, \quad \Lambda_X = \mathrm{diag}(\lambda_{X,1}, \dots, \lambda_{X,n}),$$

$$\varphi = V\Lambda_Y^{1/2}, \quad V = [v_1, \dots, v_n] = \begin{bmatrix} v_{11} & \cdots & v_{1n} \\ \vdots & \ddots & \vdots \\ v_{n1} & \cdots & v_{nn} \end{bmatrix}, \quad \Lambda_Y = \mathrm{diag}(\lambda_{Y,1}, \dots, \lambda_{Y,n}),$$

so that

$$\phi_{ij} = \sqrt{\lambda_{X,j}}\, u_{ij}, \qquad \varphi_{ik} = \sqrt{\lambda_{Y,k}}\, v_{ik}, \qquad i, j, k = 1, \dots, n.$$

Thus, arrays `S` and `T` (lines 23, 25) store ϕ_{ij} and φ_{ik}, respectively; in `W` (line 31), the two-dimensional index (j, k) is flattened to one dimension so that $w_{i,j,k} = \phi_{ij}\varphi_{ik}$ is stored as `W[i, r*(i-1)+j]`.

The eigenvalues decay roughly exponentially, and computing eigenvalues of an $n^2 \times n^2$ matrix is often infeasible. Therefore, we select the top r eigenvalues $\lambda_{X,1}, \dots, \lambda_{X,n}$ and $\lambda_{Y,1}, \dots, \lambda_{Y,n}$ and the corresponding eigenvectors $u_1, \dots, u_r$ and $v_1, \dots, v_r$, reducing W to size $n \times r^2$. The results are nearly unchanged compared to $r = n$.

Example 36 Let U, V, W be independent random variables taking values ± 1 with equal probability, and define X, Y, Z as follows: Generate $n = 100$ samples and test $X \perp\!\!\!\perp Y \mid Z$.

1. $X = U, Y = V, Z = W$.
2. $X = UW, Y = VW, Z = W$.
3. $X = U, Y = V, Z = UV$.
4. $X = U, Y = UV, Z = V$.

With $\lambda = 0.1$ and $r = 10$, we ran the program below. The statistic and null distribution for each case are shown in Fig. 4.4. In (a) and (b), the statement $X \perp\!\!\!\perp Y \mid Z$ is true, whereas in (c) and (d), it is false. The statistical test accepts the null in (a) and (b) and rejects it in (c) and (d). ■

```
## Kernels (Gaussian/RBF)
k.x <- function(x, y) exp(-sum((x - y)^2) / 2)
k.y <- k.x
k.z <- k.x

## Data
n <- 100
x <- rnorm(n)
y <- rnorm(n)
z <- rnorm(n)

## Kernel matrices
K.x <- K(k.x, x)
K.y <- K(k.y, y)
K.z <- K(k.z, z)

## Run KCI
lambda <- 0.001
result <- KCI.1(K.x, K.y, K.z, lambda)
eigen <- result$eigen_values
u <- result$statistics

alpha <- 0.05
```

```
result2 <- null.dist(eigen, alpha)
z <- result2$z
v <- result2$critical

## Plot
plot(
  density(z),
  xlim = c(min(z, v, u), max(z, v, u)),
  main = "KCI: Null distribution and observed statistic",
  xlab = "Statistic",
  ylab = "Density"
)
abline(v = v, col = "red", lty = 2, lwd = 2)  # critical value
abline(v = u, col = "blue", lty = 1, lwd = 2)  # observed statistic
```

If λ is too small, numerical instability may occur (e.g., negative eigenvalues), causing errors. Here, we use $\lambda = 0.001$ and $\alpha = 0.05$, the same settings as in the CRAN package `CondIndTests` (Kernel Conditional Independence Test) by Kun Zhang and collaborators [29]. Although we use the Gaussian kernel with $\sigma^2 = 1$, here, in practice, one should choose σ^2 by cross-validation or similar (Problem 36). In Chap. 5, we use parameters consistent with `CondIndTests`.

Appendix

Proof of Proposition 14

Since

$$HSIC(X, Y) = \|m_{XY}\|^2 - 2\langle m_{XY}, m_X m_Y\rangle + \|m_X m_Y\|^2,$$

we compute each term as follows:

$$\begin{aligned}
\|m_{XY}\|^2 &= \big\langle \mathbb{E}_{XY}[k_{\mathcal{X}}(X, \cdot)k_{\mathcal{Y}}(Y, \cdot)],\ \mathbb{E}_{X'Y'}[k_{\mathcal{X}}(X', \cdot)k_{\mathcal{Y}}(Y', \cdot)]\big\rangle \\
&= \mathbb{E}_{XY}\mathbb{E}_{X'Y'}\big[\langle k_{\mathcal{X}}(X, \cdot)k_{\mathcal{Y}}(Y, \cdot),\ k_{\mathcal{X}}(X', \cdot)k_{\mathcal{Y}}(Y', \cdot)\rangle\big] \\
&= \mathbb{E}_{XX'YY'}[k_{\mathcal{X}}(X, X')\, k_{\mathcal{Y}}(Y, Y')],
\end{aligned}$$

$$\begin{aligned}
\langle m_{XY}, m_X m_Y\rangle &= \big\langle \mathbb{E}_{XY}[k_{\mathcal{X}}(X, \cdot)k_{\mathcal{Y}}(Y, \cdot)],\ \mathbb{E}_{X'}[k_{\mathcal{X}}(X', \cdot)]\,\mathbb{E}_{Y'}[k_{\mathcal{Y}}(Y', \cdot)]\big\rangle \\
&= \mathbb{E}_{XY}\big\{\mathbb{E}_{X'}\big[\langle k_{\mathcal{X}}(X, \cdot)k_{\mathcal{Y}}(Y, \cdot),\ k_{\mathcal{X}}(X', \cdot)\,\mathbb{E}_{Y'}[k_{\mathcal{Y}}(Y', \cdot)]\rangle\big]\big\} \\
&= \mathbb{E}_{XY}\big\{\mathbb{E}_{X'}[k_{\mathcal{X}}(X, X')]\,\mathbb{E}_{Y'}[k_{\mathcal{Y}}(Y, Y')]\big\},
\end{aligned}$$

$$\begin{aligned}
\|m_X m_Y\|^2 &= \big\langle \mathbb{E}_X[k_{\mathcal{X}}(X, \cdot)]\,\mathbb{E}_Y[k_{\mathcal{Y}}(Y, \cdot)],\ \mathbb{E}_{X'}[k_{\mathcal{X}}(X', \cdot)]\,\mathbb{E}_{Y'}[k_{\mathcal{Y}}(Y', \cdot)]\big\rangle \\
&= \mathbb{E}_X\mathbb{E}_{X'}[k_{\mathcal{X}}(X, X')]\,\mathbb{E}_Y\mathbb{E}_{Y'}[k_{\mathcal{Y}}(Y, Y')].
\end{aligned}$$

■

Proof of Proposition 15

Using $\mathrm{tr}(HK_{\mathcal{X}}H \cdot HK_{\mathcal{Y}}H) = \mathrm{tr}(K_{\mathcal{X}}H \cdot HK_{\mathcal{Y}}H \cdot H)$ and $H^2 = H$, we have

$$\mathrm{tr}(\tilde{K}_{\mathcal{X}}\tilde{K}_{\mathcal{Y}}) = \mathrm{tr}(K_{\mathcal{X}}HK_{\mathcal{Y}}H),$$

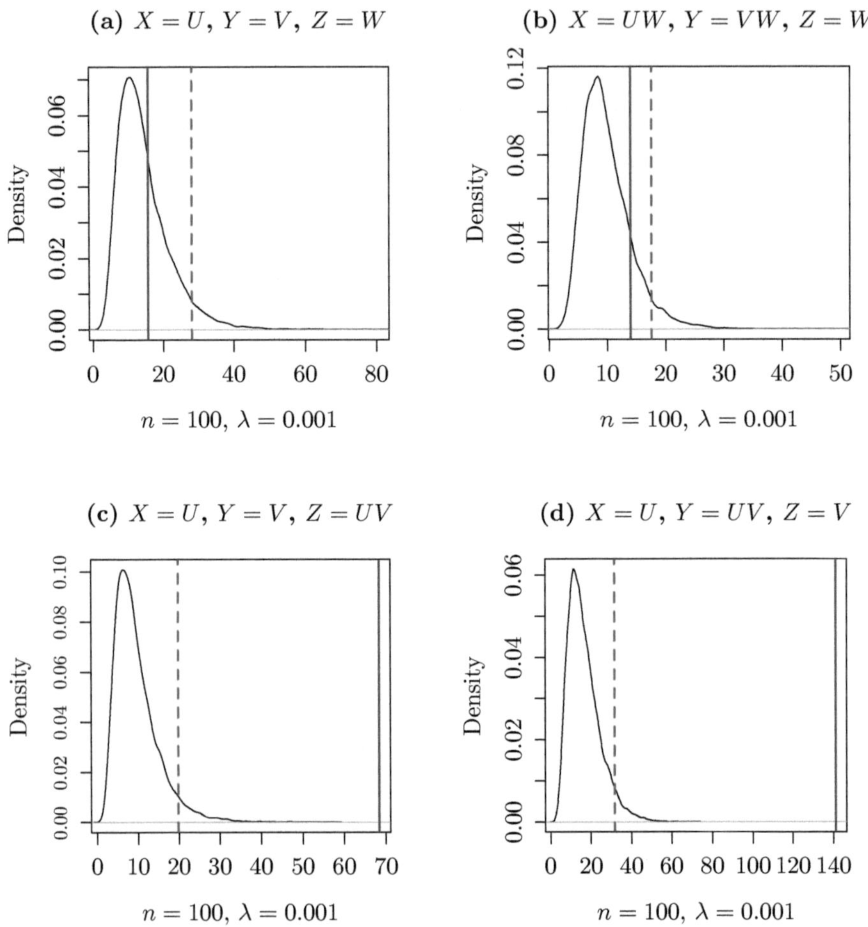

Fig. 4.4 Conditional Independence Tests for Example 36

which equals $\widehat{HSIC}$ as follows:

$$\begin{aligned}
&\mathrm{tr}(K_{\mathcal{X}} H K_{\mathcal{Y}} H) \\
&= \sum_i (K_{\mathcal{X}} H K_{\mathcal{Y}} H)_{ii} = \sum_i \sum_j (K_{\mathcal{X}} H)_{ij} (K_{\mathcal{Y}} H)_{ji} \\
&= \sum_i \sum_j \left\{ \sum_h k_{\mathcal{X}}(x_i, x_h)(\delta_{hj} - \tfrac{1}{n}) \right\} \left\{ \sum_h k_{\mathcal{Y}}(y_j, y_h)(\delta_{hi} - \tfrac{1}{n}) \right\} \\
&= \sum_i \sum_j \left\{ k_{\mathcal{X}}(x_i, x_j) k_{\mathcal{Y}}(y_i, y_j) - \frac{1}{n} k_{\mathcal{X}}(x_i, x_j) \sum_h k_{\mathcal{Y}}(y_i, y_h) \right. \\
&\quad \left. - \frac{1}{n} k_{\mathcal{Y}}(y_i, y_j) \sum_h k_{\mathcal{X}}(x_i, x_h) + \frac{1}{n^2} \sum_h k_{\mathcal{X}}(x_i, x_h) \sum_r k_{\mathcal{Y}}(y_j, y_r) \right\} \\
&= \sum_i \sum_j k_{\mathcal{X}}(x_i, x_j) k_{\mathcal{Y}}(y_i, y_j) - \frac{2}{n} \sum_i \sum_j k_{\mathcal{X}}(x_i, x_j) \sum_h k_{\mathcal{Y}}(y_i, y_h) \\
&\quad + \frac{1}{n^2} \sum_i \sum_h k_{\mathcal{X}}(x_i, x_h) \sum_j \sum_r k_{\mathcal{Y}}(y_j, y_r).
\end{aligned}$$

∎

Proofs of Propositions 16 and 17

Let $\phi = (\phi_{ij})$ and $\varphi = (\varphi_{ik})$, and set $w_{i,j,k} := \phi_{ij}\varphi_{ik}$. Then

$$\begin{aligned}
\mathrm{tr}\left(\tilde{K}_{\mathcal{X}\mathcal{Z}|\mathcal{Z}} \tilde{K}_{\mathcal{Y}\mathcal{Z}|\mathcal{Z}}\right) &= \mathrm{tr}(\phi\phi^\top \varphi\varphi^\top) = \mathrm{tr}(\phi^\top \varphi \varphi^\top \phi) = \mathrm{tr}\big((\phi^\top \varphi)(\phi^\top \varphi)^\top\big) \\
&= \|\phi^\top \varphi\|_F^2 = \sum_{j=1}^n \sum_{k=1}^n \left(\sum_{i=1}^n \phi_{ij}\varphi_{ik} \right)^2 \\
&= \sum_{j=1}^n \sum_{k=1}^n \left(\sum_{i=1}^n w_{i,j,k} \right)^2,
\end{aligned}$$

where $\|A\|_F$ denotes the Frobenius norm $\sqrt{\sum_i \sum_j a_{ij}^2}$. We used $\mathrm{tr}(AB) = \mathrm{tr}(BA)$ when both products are defined and $\mathrm{tr}(AA^\top) = \|A\|_F^2$ (Problem 41). Hence set

$$S_n := \frac{1}{\sqrt{n}} \sum_{i=1}^{n} w_i \in \mathbb{R}^{n^2}, \qquad T = \|S_n\|^2.$$

We use the following lemma:

Lemma 1 (Daudin 1980 [6]) *Let $f : \mathcal{X} \times \mathcal{Z} \to \mathbb{R}$ and $g : \mathcal{Y} \times \mathcal{Z} \to \mathbb{R}$ be square-integrable with $\mathbb{E}_X[f \mid Z] = 0$ and $\mathbb{E}_Y[g \mid Z] = 0$. Then, almost surely in Z,*

$$\mathbb{E}_{XY}[f(X, Z)g(Y, Z)] = 0$$

if and only if $X \perp\!\!\!\perp Y \mid Z$.[7]

Since $\sum_{i=1}^{n} \phi_{ij} = \sum_{i=1}^{n} \varphi_{ik} = 0$ (Problem 36), under the null $X \perp\!\!\!\perp Y \mid Z$ the vector $w_i = (\phi_{ij}\varphi_{ik})$ has mean zero by Lemma 1. By the multivariate central limit theorem, S_n is asymptotically normal with mean zero, and by the law of large numbers, $\frac{1}{n} w w^\top$ converges to the covariance matrix $\Sigma \in \mathbb{R}^{n^2 \times n^2}$. Since T is the sum of squares of the components of S_n, left-multiplying by an orthogonal matrix $P \in \mathbb{R}^{n^2 \times n^2}$ does not change T. We can diagonalize Σ to $\mathrm{diag}(\lambda_1/n, \ldots, \lambda_{n^2}/n)$. Therefore, under the null and for large n,

$$T \sim \frac{1}{n} \sum_{k=1}^{n^2} \lambda_k Z_k^2,$$

with independent $Z_k \sim N(0, 1)$.

Finally, we show that Proposition 16 is a special case of Proposition 17. The eigenvalues of the null distribution are products $\lambda_{X,j}\lambda_{Y,k}$. With $\phi_{ij} = \sqrt{\lambda_{X,j}} u_{ij}$ and $\varphi_{ik} = \sqrt{\lambda_{Y,k}} v_{ik}$, the $((j, k), (j', k'))$ entry of Σ is

$$\frac{1}{n} \sum_{i=1}^{n} \sqrt{\lambda_{X,j}\lambda_{X,j'}} \sqrt{\lambda_{Y,k}\lambda_{Y,k'}}\, u_{ij}u_{ij'}\, v_{ik}v_{ik'}.$$

Under the null and asymptotically,

$$\frac{1}{n} \sum_{i=1}^{n} \sqrt{\lambda_{X,j}\lambda_{X,j'}}\, u_{ij}u_{ij'} \cdot \frac{1}{n} \sum_{i'=1}^{n} \sqrt{\lambda_{Y,k}\lambda_{Y,k'}}\, v_{i'k}v_{i'k'}$$

so that diagonal entries converge to $\lambda_{X,j}\lambda_{Y,k}$ and off-diagonals to 0, using $\sum_{i=1}^{n} u_{ij}u_{ij'} = \delta_{jj'}$ and $\sum_{i=1}^{n} v_{ik}v_{ik'} = \delta_{kk'}$. ■

[7] It suffices if one of f or g depends only on X or Y with zero mean.

Problems 30–43

30. I would like to implement the Nadaraya–Watson estimator and obtain an output like Fig. 4.1. Construct (i) the Epanechnikov kernel function `D`, (ii) a function `K` that returns the value of the Gaussian kernel for $x, y \in \mathbb{R}^n$ and $\sigma^2 > 0$, and (iii) a function `f` that predicts y_* at a new value $x_* \in \mathbb{R}$ with $\sigma^2 > 0$ via the Nadaraya–Watson estimator. Apply them to the program below and verify its behavior.

```
n <- 250
x <- 2*rnorm(n)
y <- sin(2*pi*x) + rnorm(n)/4  ## data generation
D <- ## define the function
k <- ## define the function
f <- ## define the function
# plotting setup
plot(seq(-3, 3, length = 10), seq(-2, 3, length = 10), type = "n",
     xlab = "x", ylab = "y")
points(x, y, pch = 16, cex = 0.5)
xx <- seq(-3, 3, 0.05)
lambdas <- c(0.05, 0.35, 0.50)
colors <- c("green", "blue", "red")
# plot estimates
for (i in 1:3) {
  yy <- sapply(xx, function(z) f(z, lambdas[i]))
  lines(xx, yy, col = colors[i], lwd = 2)
}
```

31. Show that at least one eigenvalue of the Gram matrix K_λ in Example 29 is negative, and hence K_λ is not positive semidefinite. [Hint] For a symmetric matrix $A \in \mathbb{R}^{N\times N}$, A is positive semidefinite if and only if all its eigenvalues are nonnegative.
32. Show that the Gaussian kernel and the Cauchy kernel are, respectively, characteristic functions of the corresponding Cauchy distribution. Also, for $\mathcal{X} = \mathbb{R}$, write R functions that output the values of the Gaussian kernel with $\sigma^2 = 1$ and the Laplace kernel with $\beta = 1$.
33. Show that the inner product of a finite-dimensional Euclidean space defines a kernel.
34. Let k be a positive definite kernel. Show that the set $H_0 = \{f = \sum_{i=1}^N \alpha_i k(x_i, \cdot) : N \geq 1,\ x_1, \ldots, x_N \in \mathcal{X},\ \alpha_1, \ldots, \alpha_N \in \mathbb{R}\}$ satisfies (4.1). Also show that the inner product $\sum_{i=1}^N \sum_{j=1}^N \alpha_i \beta_j\, k(x_i, x_j)$ for $f = \sum_{i=1}^N \alpha_i k(x_i, \cdot)$ and $g = \sum_{j=1}^N \beta_j k(x_j, \cdot) \in H_0$ satisfies (4.2).
35. Show that L^2 in Example 32 is a linear space. Also, state how the norm $\|f\|$ of each element f is defined.

36. Using the fact that each row sum and each column sum of the centered Gram matrix $\tilde{K}$ equals 0, show that $\sum_{i=1}^{n} \phi_{i,j} = \sum_{i=1}^{n} \varphi_{i,j} = 0$. [Hint] Derive $1^\top R_\mathcal{Z} V_{\mathcal{X}\mathcal{Z}} = 1^\top V_{\mathcal{X}\mathcal{Z}} = 0$.
37. For the datasets below, display the relationship between the null distribution and the test statistic in the same format as the figure in Example 35.

```
x <- rnorm(n); y <- rnorm(n)
x <- rnorm(n); y <- rnorm(n) + x
```

38. Write $f \in H$ as the sum of a linear combination of $k(x_i, \cdot)$, $i = 1, \dots, n$, namely $\sum_{i=1}^{n} \alpha_i k(x_i, \cdot)$, and a function $f_\perp$ orthogonal to that span. Then:

    (a) Show that $\|f\|^2 = \left\|\sum_{i=1}^{n} \alpha_i k(x_i, \cdot)\right\|^2 + \|f_\perp\|^2$.
    (b) Show that the minimizer $f \in H$ of (4.8) can be expressed as a linear combination of $k(x_i, \cdot)$ for $i = 1, \dots, n$.

39. Using the same data as in Example 36 (all four cases), change the parameters to $\lambda = 0.01$ and $r = 3, 20$, run the procedure, and check whether the results change.
40. In Example 36, let each random variable follow the standard normal distribution, and define

    (a) $X = U, Y = V, Z = W$,
    (b) $X = U + W, Y = V + W, Z = W$,
    (c) $X = U, Y = V, Z = U + V$,
    (d) $X = U, Y = U + V, Z = V$.

    For each case, display the test statistic and the null distribution in the style of Fig. 4.4.
41. Show that $\mathrm{tr}(AB) = \mathrm{tr}(BA)$ whenever both products AB and BA are defined, and that for a matrix $A = (a_{ij})$, $\mathrm{tr}(AA^\top) = \|A\|_F^2$.
42. Unconditional independence corresponds to Z taking a fixed value. Suppose $k_\mathcal{Z}(z, z')$ is a constant $a > 0$ for all $z, z' \in \mathcal{Z}$. Show the following:

    (a) $R_\mathcal{Z}$ is the identity matrix.
    (b) Both T and $\frac{1}{n}\sum_{k=1}^{n^2} \lambda_k Z_k^2$ are multiplied by a^2.

43. In Example 9, we presented a case where the correlation between X and Y is zero, but they are not independent. In such cases, nonindependence cannot be detected by the correlation coefficient. Generate $n = 100$ random samples of the form (x_i, x_i^2), $i = 1, \dots, n$ and test independence using HSIC (significance level $\alpha = 0.01$). Furthermore, let Z be a binary random variable taking values 0 and 1 with equal probability. When $Z = 0$, set $Y = X^2$; when $Z = 1$, set $X = Y^2$. Generate $n = 100$ random triples (x_i, y_i, z_i) and test conditional independence using KCI.

Chapter 5
The PC Algorithm

In this chapter, we explain the Peter–Clark (PC) algorithm, which constructs a directed acyclic graph (DAG) based on tests of conditional independence. The algorithm enables efficient structure estimation, particularly for high-dimensional data.

We begin by describing the *faithfulness* assumption that underpins the theoretical validity of the PC algorithm: every conditional independence in the distribution can be represented by d-separation in some DAG. We then gain an overview using the CRAN package `pcalg`.

Next, we present the first stage of the algorithm that decides the presence or absence of edges using general conditional independence tests such as Kernel-based conditional independence (KCI). Along with R implementations, we detail how the undirected skeleton is built.

We then explain the second stage, which orients edges based on rules that eliminate colliders and directed cycles, and we clarify how a completed partially directed acyclic graph (CPDAG) is produced. When full orientation is impossible, the resulting structure allows partially directed edges (a CPDAG).

Finally, we apply the procedures discussed in this chapter to the Boston dataset.

(**Note.** The programs in Chap. 5 assume that the programs in Chap. 4 have been run.)

5.1 Overview of the PC Algorithm

The PC algorithm constructs a graphical model (a DAG) to infer causal relations by repeatedly testing conditional independence between variables. Starting from a complete undirected graph, it removes unnecessary edges via conditional independence tests and finally assigns directions to edges in line with causal constraints.

J. Suzuki, *Graphical Models and Causal Discovery with R*,
https://doi.org/10.1007/978-981-95-4267-3_5

It is especially effective for high-dimensional data and is widely used in Bayesian networks and causal inference.

The name *PC algorithm* (Peter and Clark algorithm) [28] derives from its developers Peter Spirtes and Clark Glymour, who proposed it in the line of research on statistical causal discovery.

The PC algorithm assumes that the conditional independences among the random variables are *faithful*. Below, we write conditional independence among subsets of variables $X_1, \ldots, X_p$ as $X_S \perp\!\!\!\perp X_T \mid X_U$, where S, T, U are pairwise disjoint subsets of $\{1, \ldots, p\}$. We also write separation and d-separation as $S \perp\!\!\!\perp_G T \mid U$. Note that in general, there may be multiple triples (S, T, U) that satisfy a given relation. We denote this by

$$\bigwedge_{(S,T,U)} (X_S \perp\!\!\!\perp X_T \mid X_U), \qquad \bigwedge_{(S,T,U)} (S \perp\!\!\!\perp_G T \mid U).$$

As noted in Chap. 3, whether G is an undirected graph (Markov network, MN) or a DAG (Bayesian network, BN), it is necessary that

$$\bigwedge_{(S,T,U)} (S \perp\!\!\!\perp_G T \mid U) \Longrightarrow \bigwedge_{(S,T,U)} (X_S \perp\!\!\!\perp X_T \mid X_U). \tag{5.1}$$

However, the converse

$$\bigwedge_{(S,T,U)} (S \perp\!\!\!\perp_G T \mid U) \Longleftarrow \bigwedge_{(S,T,U)} (X_S \perp\!\!\!\perp X_T \mid X_U) \tag{5.2}$$

does not necessarily hold (Examples 25 and 26). Whether (5.2) holds depends on the specific random variables $X_1, \ldots, X_p$. If there exists G for which both (5.1) and (5.2) hold, then $X_1, \ldots, X_p$ are said to be *faithful*. More precisely, if G is an undirected graph, they are faithful with respect to undirected graphs; if G is a DAG, they are faithful with respect to DAGs.

Example 37 Roll two dice A and B. Let $X_1 \in \{1, 2, 3, 4, 5, 6\}$ be the outcome of die A; let $X_2 = 1$ if the outcome of B is odd and $X_2 = 0$ otherwise; and let $X_3 = 0$ if the outcome of B is at most 3 and $X_3 = 1$ otherwise. Figure 3.6b represents a BN for such X_1, X_2, X_3. Indeed, we have $X_1 \perp\!\!\!\perp X_2$, $X_1 \perp\!\!\!\perp X_3$, and $X_1 \perp\!\!\!\perp \{X_2, X_3\}$. Since also $1 \perp\!\!\!\perp_G 2$, $1 \perp\!\!\!\perp_G 3$, and $1 \perp\!\!\!\perp_G \{2, 3\}$ hold, X_1, X_2, X_3 are faithful with respect to DAGs. Similarly, Fig. 3.7b shows that they are faithful with respect to undirected graphs as well.

However, if we keep X_1 as the outcome of A, take X_2 as the outcome of B, and let X_3 be the sum of the outcomes of A and B, then the only non-trivial independence is $X_1 \perp\!\!\!\perp X_2$. The BN representation is Fig. 3.6j, whereas among MNs, only Fig. 3.7h represents the relations. In this sense, X_1, X_2, X_3 are faithful with respect to DAGs, but not with respect to undirected graphs. ■

From (3.5), for conditional independence, we always have

$$X_S \perp\!\!\!\perp X_T \mid X_U \Longrightarrow \bigwedge_{i \in S,\, j \in T} (X_i \perp\!\!\!\perp X_j \mid X_U). \tag{5.3}$$

The converse of (5.3) does not hold in general.

Example 38 Let X_1, X_2 be independent Rademacher variables taking ± 1 with equal probability, and set $X_3 = X_1 X_2$. Then $X_1 \perp\!\!\!\perp X_2$ and $X_1 \perp\!\!\!\perp X_3$ hold, but $X_1 \perp\!\!\!\perp \{X_2, X_3\}$ does *not* hold, because X_2 and X_3 together determine X_1. Hence,

$$X_1 \perp\!\!\!\perp X_2,\ X_1 \perp\!\!\!\perp X_3 \not\Longrightarrow X_1 \perp\!\!\!\perp \{X_2, X_3\}.$$

This means

$$1 \perp\!\!\!\perp_G \{2, 3\} \not\Longrightarrow X_1 \perp\!\!\!\perp X_{2,3},$$

so (5.1) holds, whereas (5.2) fails in this example. In fact, as shown in (3.8) for separation and d-separation, for each (S, T, U), we have

$$\bigwedge_{i \in S,\, j \in T} (i \perp\!\!\!\perp_G j \mid U) \Longleftrightarrow S \perp\!\!\!\perp_G T \mid U. \tag{5.4}$$

■

Thus, knowing the truth values of $i \perp\!\!\!\perp_G j \mid U$ for all $(i, j) \in S \times T$ determines that of $S \perp\!\!\!\perp_G T \mid U$.

Using this fact, we obtain the following key proposition for the PC algorithm.

Proposition 18 *If $X_1, \ldots, X_p$ are faithful with respect to an undirected graph or a DAG, then for conditional independence, we have*

$$\bigwedge_{i \in S,\, j \in T} (X_i \perp\!\!\!\perp X_j \mid X_U) \Longrightarrow X_S \perp\!\!\!\perp X_T \mid X_U. \tag{5.5}$$

That is, if we know the truth values of $X_i \perp\!\!\!\perp X_j \mid X_U$ for all $(i, j) \in S \times T$, then we know that of $X_S \perp\!\!\!\perp X_T \mid X_U$.

Proof Assuming faithfulness, i.e., (5.2), apply (5.2), (5.4), and (5.1) in this order to obtain

$$\begin{aligned}
\bigwedge_{i \in S,\, j \in T} (X_i \perp\!\!\!\perp X_j \mid X_U) &\Longrightarrow \bigwedge_{i \in S,\, j \in T} (i \perp\!\!\!\perp_G j \mid U) \\
&\Longrightarrow S \perp\!\!\!\perp_G T \mid U \\
&\Longrightarrow X_S \perp\!\!\!\perp X_T \mid X_U,
\end{aligned}$$

which proves (5.5). ■

Example 39 Suppose (X, Y, Z) is trivariate normal. If $X \perp\!\!\!\perp Z$ and $Y \perp\!\!\!\perp Z$ hold simultaneously, then the covariance matrix has the form

$$\begin{bmatrix} * & * & 0 \\ * & * & 0 \\ 0 & 0 & * \end{bmatrix}$$

(where "$*$" denotes a nonzero entry), and the joint density factorizes as

$$f_{XYZ}(x, y, z) = f_{XY}(x, y)\, f_Z(z)$$

(Prop. 1). Hence, $\{X, Y\} \perp\!\!\!\perp Z$ holds. The same argument extends to a general multivariate normal distribution, so (5.5) holds in that setting (Problem 44). ■

Under the faithfulness assumption (in essence, under (5.5)), the PC algorithm restricts conditional independence tests to the form $X_i \perp\!\!\!\perp X_j \mid X_U$ and thereby efficiently uncovers the structure.

In other words, we ensure that for each $(i, j) \in S \times T$,

$$i \perp\!\!\!\perp_G j \mid U \implies X_i \perp\!\!\!\perp X_j \mid X_U.$$

Concretely, for each pair (i, j), we seek a maximal conditioning set U for which $X_i \perp\!\!\!\perp X_j \mid X_U$ holds, and then use that information to orient edges. In outline:

- Perform conditional independence tests to obtain the *skeleton* (an undirected graph) from data (Sect. 5.3).
- Decide the directions of edges in the skeleton (Sect. 5.4).

The resulting BN is generated in two stages. Some directions cannot be decided due to Markov equivalence. In such cases, Chap. 6's LiNGAM can be effective.

A well-known implementation of the PC algorithm is the CRAN package `pcalg` [10]. Note that the original PC algorithm can suffer from order-dependence and long runtimes, so a *stable* version is now commonly used. As of March 2025, KCI (treated in Chap. 4) is not yet integrated as a conditional independence test (e.g., tests based on covariance matrices are available).

Installation:

```
install.packages("tidyverse")
install.packages("ggExtra")
install.packages("BiocManager")
BiocManager::install(c("graph", "RBGL", "Rgraphviz"))
install.packages("pcalg")
```

In particular, `graph`, `RBGL`, and `Rgraphviz` must be installed via `BiocManager`, so install `BiocManager` first.

For execution, use `skeleton` to obtain only the skeleton, or `pc` to obtain the full result (a DAG or a CPDAG with equivalent information). For example:

```
library(pcalg)
data(gmG)
suffStat <- list(C = cor(gmG$x), n = nrow(gmG$x))
skeleton.fit <- pcalg::skeleton(suffStat, indepTest = gaussCItest,
                                p = ncol(gmG$x), alpha = 0.01)
pc.fit <- pc(suffStat, indepTest = gaussCItest,
             p = ncol(gmG$x), alpha = 0.01)
par(mfrow = c(1,3)) # arrange three plots horizontally
plot(gmG$g, main = "True DAG")
plot(skeleton.fit, main = "Estimated Skeleton")
plot(pc.fit, main = "Estimated CPDAG")
```

For `skeleton` and `pc`, one must specify the sufficient statistics `suffStat`, the independence test `indepTest`, the number of variables `p`, and the significance level `alpha`. The sufficient statistics depends on the test: for continuous variables, only a Gaussian-based test is provided, with `suffStat` being a correlation matrix and independence implied by zero correlation. In the code, `gmG` contains artificial data from a DAG: `gmG$x` are samples generated from the DAG with certain parameters, and `gmG$g` is the true DAG. Figure 5.1 shows the following: left, the true DAG; middle, the estimated skeleton; and right, the estimated CPDAG (a graph encoding both possible directions when orientation is not unique).

`gmG` is artificial data distributed with `pcalg`.

In the remainder of this chapter, we focus on the essence of how the PC algorithm works.

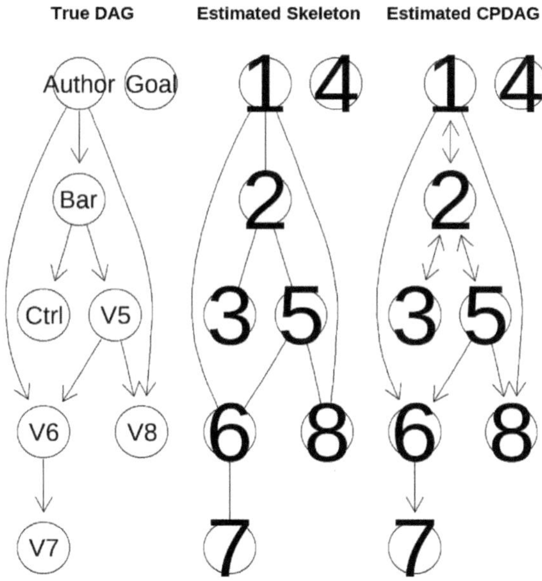

Fig. 5.1 Graphs obtained by applying `pcalg` to the dataset gmG: left, the true DAG; middle, the estimated skeleton; right, the estimated CPDAG

5.2 Conditional Independence Testing in the PC Algorithm

Chapter 4 studied tests of the form $X \perp\!\!\!\perp Y \mid Z$ (a single conditioning variable). In the PC algorithm, we keep X, Y as before but must generalize to multiple conditioning variables Z.

In short, for testing $X \perp\!\!\!\perp Y \mid \{Z, W\}$, suppose we have Gram matrices K_X, K_Y, K_Z, K_W. Let K_{ZW} denote the Hadamard (elementwise) product of K_Z and K_W. Then, wherever we previously formed a test statistic for $X \perp\!\!\!\perp Y \mid Z$ using K_X, K_Y, K_Z, we now form it for $X \perp\!\!\!\perp Y \mid \{Z, W\}$ using K_X, K_Y, K_{ZW}.

If we prepare kernels $k^{(1)}, \ldots, k^{(p)}$ for p variables and observe samples $x^{(1)}, \ldots, x^{(p)} \in \mathbb{R}^n$, we obtain p Gram matrices

$$K^{(1)} = \left(k^{(1)}(x_s^{(1)}, x_t^{(1)})\right)_{s,t=1}^n, \ \ldots, \ K^{(p)} = \left(k^{(p)}(x_s^{(p)}, x_t^{(p)})\right)_{s,t=1}^n.$$

To test $X_i \perp\!\!\!\perp X_j \mid X_S$ (where S is an index set), we use $K^{(i)}$, $K^{(j)}$, and the Hadamard product over $K^{(h)}$ for $h \in S$. In this sense, the R function `KCI.1` defined in Chap. 4 can be used as is. A helper for Hadamard products in R is:

```
# Compute the Hadamard product of Gram matrices indexed by S
K.merge <- function(S, K.list) {
  n <- nrow(K.list[[1]])  # size of the matrices

  # Initialize K as an n x n matrix of ones (not the identity)
  K <- matrix(1, n, n)

  # Elementwise (Hadamard) product over S
  for (h in S) {
    K <- K * K.list[[h]]
  }
  return(K)
}
```

We initialize with an all-ones matrix (not the identity) and use * (elementwise product) rather than %*% (matrix product).

Using the previously defined functions, we write the following R function. The p Gram matrices $K^{(1)}, \ldots, K^{(p)}$ are stored in the list `K.list`.

```
CI.PC <- function(i, j, S, K.list, alpha, lambda) {
  n <- nrow(K.list[[1]])
  m <- length(S)

  # Use HSIC (if m=0) or KCI (if m>0)
  if (m == 0) {
    result <- HSIC.2(K.list[[i]], K.list[[j]])
  } else {
    KK <- K.merge(S, K.list)
    result <- KCI.1(K.list[[i]], K.list[[j]], KK, lambda)
  }

```

```
  statistics <- result$statistics
  eigen <- result$eigen_values

  # Critical value at level alpha from the null distribution
  critical <- null.dist(eigen, alpha)$critical

  # Accept independence if statistic < critical
  True.False <- (statistics < critical)

  # (Optional) logging
  print(c(i, j))
  print(S)
  print(True.False)

  return(True.False)
}
```

We can now obtain concrete test outcomes as follows:

```
# Settings
p <- 4
n <- 100

# Sample generation
z <- rnorm(n)
w <- rnorm(n)
x <- z + rnorm(n)       # x depends on z
y <- w + rnorm(n)       # y depends on w

X <- cbind(x, y, z, w)

# Gaussian kernel
gaussian_kernel <- function(x, y) exp(-sum((x - y)^2) / 2)

# Gram matrices for each variable
KK <- list()
for (j in 1:p) {
  KK[[j]] <- K(gaussian_kernel, X[, j])
}

# Test X1   X2 | {X3, X4}
result <- CI.PC(1, 2, c(3, 4), KK, 0.05, 0.1)
print(result)
```

Example output:

```
[1] TRUE
# If we change y <- w + rnorm(n) to y <- w + rnorm(n) + x
# and re-run, we may obtain
[1] FALSE
```

Thus, we have established a method for conditional independence testing. However, it is not obvious in what order to test among p variables so that the desired undirected graph is obtained.

5.3 Constructing the Skeleton

The following procedure implements the skeleton-learning stage of the PC algorithm in R. It starts from the complete graph and deletes edges via conditional independence tests.

We prepare `adj`, a matrix storing adjacency between variables, initialized to `TRUE` everywhere off-diagonal and `FALSE` on the diagonal. We also prepare a 3D array `s` to store separating-set information (`s[i,j,k]` records whether k belongs to some maximal separating set S that makes $X_i \perp\!\!\!\perp X_j \mid X_S$), initialized to `FALSE`.

Inside the loop, we consider each pair `i`, `j` and test independence under conditioning sets `S` of increasing size. If `m=0`, we test marginal independence; if `m>0`, we generate all subsets of size `m` from the current neighbors using `combn`, and test $X_i \perp\!\!\!\perp X_j \mid X_S$ for each. When independence is accepted, the corresponding edge is deleted, and we record the conditioning variables in `s`.

The loop terminates when no more deletions occur for a given `m`, and we then increase `m` by one.

```r
skeleton <- function(K.list, alpha, lambda) {
  p <- length(K.list)
  m <- 0
  repeat {
    done <- TRUE
    for (i in 1:(p - 1)) {
      for (j in (i + 1):p) {
        if (adj[i, j]) {
          S <- setdiff(which(adj[i, ]), c(i, j))
          if (length(S) >= m) {
            if (m == 0) {
              if (CI.PC(i, j, NULL, K.list, alpha, lambda)) {
                adj[i, j] <- FALSE
                adj[j, i] <- FALSE
                done <- FALSE
              }
            } else {
              mat <- combn(S, m)
              for (h in 1:ncol(mat)) {
                cond_set <- mat[, h]
                if (CI.PC(i, j, cond_set, K.list, alpha, lambda)) {
                  adj[i, j] <- FALSE
                  adj[j, i] <- FALSE
                  for (k in cond_set) {
                    s[i, j, k] <- TRUE
                    s[j, i, k] <- TRUE
                  }
                  done <- FALSE
                  break
                }
              }
            }
          }
```

```
        }
      }
    }
    if (done) break
    m <- m + 1
  }
  return(list(adj = adj, s = s))
}
```

Example 40 The following code generates data according to the left panel of Fig. 5.2. The resulting $n \times p$ matrix is stored in `X`. Using `skeleton` on this data yields the skeleton in the right panel of Fig. 5.2.

```
# Data generation
n <- 200
p <- 4
y <- rnorm(n)
z <- rnorm(n)
x <- z + rbinom(n, 1, 0.25)
w <- y + z + rbinom(n, 1, 0.25)
X <- cbind(x, y, z, w)
```

■

We now compute Gram matrices and estimate the skeleton. Finally, we build an undirected graph from the adjacency matrix and plot it using `igraph`.

```
library(igraph)
# Gaussian (RBF) kernels for each variable
k <- list()
for (j in 1:p) {
  k[[j]] <- function(x, y) exp(-sum((x - y)^2) / 2)
}

X <- scale(X)

# Gram matrices
KK <- list()
for (j in 1:p) {
  KK[[j]] <- K(k[[j]], X[, j])
}

# Global state: adj and s
adj <- array(TRUE, dim = c(p, p))
for (j in 1:p) adj[j, j] <- FALSE
s <- array(FALSE, dim = c(p, p, p))

alpha <- 0.05
lambda <- 10^(-3)
skeleton(KK, alpha, lambda)

g <- graph_from_adjacency_matrix(adj, mode = "undirected")
plot(g)
```

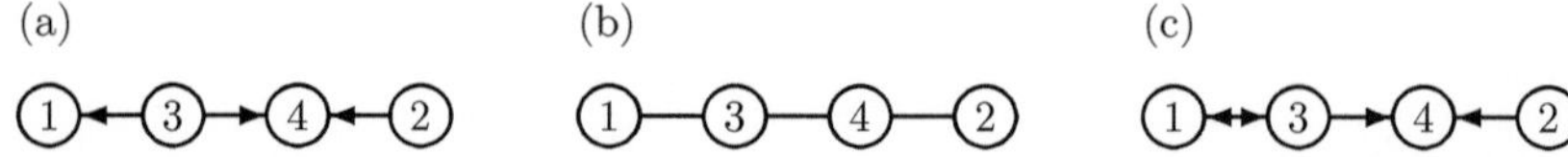

Fig. 5.2 (**a**) The dependency generating the data, (**b**) the skeleton produced by `skeleton`, and (**c**) an ordering of the edges of the skeleton with the collider $3 \to 4 \leftarrow 2$ detected. The order of $1 \leftrightarrow 3$ can be chosen arbitrarily

5.4 Orienting the Skeleton

In the second stage of the PC algorithm, we orient the edges of the skeleton.

Below, we do *not* update the truth values in the array `s[i,j,k]`. We only change entries of `adj[i,j]` from `TRUE` to `FALSE` as we prune one of the two possible directions.

Section 5.3 described the first stage (skeleton). The second stage orients edges.

Part 1 (Collider detection) From `skeleton`, we obtain `adj` with

$$\texttt{adj[i,j]} \iff \texttt{adj[j,i]}$$

(i.e., both directions are initially possible). Using the following rule (rule 0), and assuming i and k are not adjacent, we remove one of the two directions:

$$\texttt{adj[i,j]},\ \texttt{adj[k,j]},\ \text{and}\ \neg\texttt{s[i,j,k]}$$
$$\implies \texttt{adj[j,i]} \leftarrow \texttt{FALSE},\ \texttt{adj[j,k]} \leftarrow \texttt{FALSE}.$$

If $X_i \perp\!\!\!\perp X_j \mid X_S$ with $k \in S$, then `s[i,j,k] = TRUE` is required. Thus, `s[i,j,k] = FALSE` indicates a collider. Assuming all colliders are detected in Part 1, Part 2 adds further orientations (Fig. 5.3).

Part 2 (Four orientation rules)

- **Rule 1(a)** With i and k nonadjacent, if $i \to j$ and $j \leftrightarrow k$, then $j \leftarrow k$ would create a collider at j—contradicting Part 1—so we must have $j \to k$.
- **Rule 1(b)** If $i \to j \to k$ and $i \leftrightarrow k$, then $i \leftarrow k$ forms a directed cycle; thus $i \to k$.
- **Rule 1(c)** With k and l nonadjacent, if $k \to j \leftarrow l$ and we have $i \leftrightarrow j$, $i \leftrightarrow k$, $i \leftrightarrow l$, then $i \leftarrow j$ would force $i \leftarrow k$ and $i \leftarrow l$ to avoid cycles, introducing a new collider; hence $i \to j$.
- **Rule 1(d)** With k and j nonadjacent, if $k \to l \to j$ and $i \leftrightarrow k$, $i \leftrightarrow j$, then $i \to j$ is required; otherwise, a cycle or a new collider appears.

The following R function implements these orientations. Each entry of `adj` initially has both directions set to `TRUE`; we encode direction by setting exactly one of the two to `FALSE`.

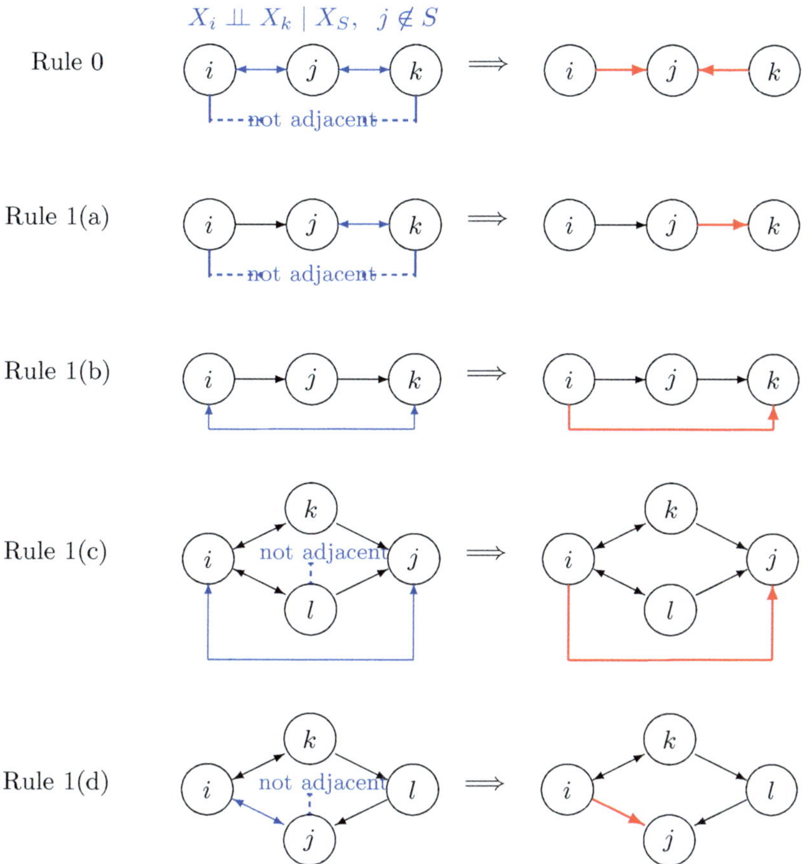

Fig. 5.3 Rule 0: If $i \perp\!\!\!\perp j \mid S(i, j)$ does not hold for any $S(i, j)$ with $k \in S(i, j)$, then $i \to j \leftarrow k$ (a collider). Rule 1(a): Since all colliders have been found in Part 1, no new collider may be introduced. Rule 1(b): Because the graph must be acyclic, directed cycles are forbidden. Rule 1(c): With $k \not\leftrightarrow l$, if $k \to j \leftarrow l$ and $i \leftrightarrow j$, $i \leftrightarrow k$, $i \leftrightarrow l$, then $i \leftarrow j$ would force $i \leftarrow k$ and $i \leftarrow l$ to avoid cycles, creating a new collider; hence, $i \to j$. Rule 1(d): With $k \not\leftrightarrow j$, if $k \to l \to j$ and $i \leftrightarrow k$, $i \leftrightarrow j$, then $i \to j$ is required; otherwise, a cycle or a new collider is created

```
Post_PC <- function(adj, s) {
  p <- nrow(adj)

  any  <- function(i, j) adj[i, j] | adj[j, i]
  no   <- function(i, j) !any(i, j)

  # Rule 0: collider detection
  for (i in 1:p) {
    for (j in setdiff(1:p, i)) {
      for (k in setdiff(1:p, c(i, j))) {
        if (no(i, k) && !s[i, k, j] && adj[i, j] && adj[k, j]) {
```

```
          adj[j, i] <- FALSE
          adj[j, k] <- FALSE
        }
      }
    }
  }

  # Rule 1(a): i->j and j<->k with i and k nonadjacent => j->k
  for (i in 1:p) {
    for (j in setdiff(1:p, i)) {
      for (k in setdiff(1:p, c(i, j))) {
        if (adj[i, j] && adj[j, k] && no(i, k)) {
          adj[k, j] <- FALSE
        }
      }
    }
  }

  # Rule 1(b): i->j->k and i<->k => i->k (avoid cycles)
  for (i in 1:p) {
    for (j in setdiff(1:p, i)) {
      for (k in setdiff(1:p, c(i, j))) {
        if (adj[i, j] && adj[j, k] && adj[i, k]) {
          adj[k, i] <- FALSE
        }
      }
    }
  }

  # Rule 1(c): collider extension (pattern with k->j<-l and i<->{j,k,l})
  for (j in 1:p) {
    for (k in setdiff(1:p, j)) {
      for (l in setdiff(1:p, c(j, k))) {
        for (i in setdiff(1:p, c(j, k, l))) {
          if (adj[k, j] && adj[l, j] && adj[i, j] && any(i, k) && any(i, l)) {
            adj[j, i] <- FALSE
          }
        }
      }
    }
  }

  # Rule 1(d): chain k->l->j with i<->k and i<->j => i->j
  for (j in 1:p) {
    for (k in setdiff(1:p, j)) {
      for (l in setdiff(1:p, c(j, k))) {
        for (i in setdiff(1:p, c(j, k, l))) {
          if (adj[k, l] && adj[l, j] && adj[i, j] && any(i, k)) {
            adj[j, i] <- FALSE
          }
        }
      }
    }
```

```
  }

  return(adj)
}
```

Example 41 Using Post_PC, we detect a collider by exploiting `s[i,j,k] = FALSE`. After `skeleton`, we have $1 \leftrightarrow 3 \leftrightarrow 2$, and since `s[1,2,3]` and `s[2,1,3]` are `FALSE`, rule 0 yields $1 \to 3 \leftarrow 2$.

```
p <- 3
adj <- matrix(FALSE, p, p)
adj[1,3] <- adj[3,1] <- TRUE
adj[2,3] <- adj[3,2] <- TRUE
diag(adj) <- FALSE

s <- array(TRUE, dim = c(p, p, p))   # initialize as all separations
s[1,2,3] <- s[2,1,3] <- FALSE        # but set S[1,2,3] = FALSE

adj_post <- Post_PC(adj, s)
adj_post
```

Example output:

```
FALSE  FALSE  TRUE
FALSE  FALSE  TRUE
FALSE  FALSE  FALSE
```

Even after Part 2, unoriented edges may remain. This may be due to Markov equivalence. In practice, there may be additional admissible orientations consistent with acyclicity and collider constraints; rules 1(a)–(d) are a useful set that avoid manual case-by-case search. A mixed graph with both directed and undirected edges produced by the PC algorithm is called a completed partially directed acyclic graph (CPDAG).

Example 42 We repeat Example 40 on the Boston housing dataset from the CRAN package `MASS` (Table 5.1). Since HSIC/KCI rely on kernels and cost about $O(n^3)$ time in the sample size n, we run on the first $n = 100$ (or $n = 200$) samples when the full dataset exceeds 500 samples.

```
library(igraph)
library(MASS)

# Boston data   # For speed, run on the first 100 samples (~10-15 minutes)
df <- Boston
X <- scale(as.matrix(df))
n <- 100
X <- X[1:n, ]
p <- ncol(X)

# Gaussian kernel
k <- function(x, y) exp(-sum((x - y)^2) / 2)

```

```
# Gram matrices
KK <- list()
for (j in 1:p) {
  KK[[j]] <- K(k, X[, j])
}

# Global state
adj <- matrix(TRUE, p, p)
diag(adj) <- FALSE
s <- array(FALSE, dim = c(p, p, p))

# Parameters
alpha <- 0.05
lambda <- 1e-3

# Skeleton (in-place updates of adj and s)
skeleton(KK, alpha, lambda)

# Orientation
adj <- Post_PC(adj, s)

# Visualization
g <- graph_from_adjacency_matrix(adj, mode = "directed")
plot(g)
```

At the skeleton stage, there were 64 directed edge entries (i.e., 32 undirected edges). After rule 0, there were 33 directed edge entries (only $1 \leftrightarrow 11$ remained bidirected). After applying rules 1(a), 1(b), 1(c), and 1(d), the numbers became 32, 30, 26, and 26, respectively (Fig. 5.4). In total, 93 instances of conditional independence were accepted during skeleton learning (Table 5.2).

■

Table 5.1 Boston dataset (CRAN MASS package)

Column	Variable	Meaning of variable
1	CRIM	Per capita crime rate by town
2	ZN	Proportion of residential land zoned for lots over 25,000 sq.ft.
3	INDUS	Proportion of non-retail business acres per town
4	CHAS	Charles River dummy variable (1 if tract bounds river; 0 otherwise)
5	NOX	Nitric oxide concentration (parts per 10 million)
6	RM	Average number of rooms per dwelling
7	AGE	Proportion of owner-occupied units built prior to 1940
8	DIS	Weighted distances to five Boston employment centers
9	RAD	Index of accessibility to radial highways
10	TAX	Full-value property tax rate per $10,000
11	PTRATIO	Pupil–teacher ratio by town
12	B	Proportion of Black residents by town
13	LSTAT	Percentage of lower-status population
14	MEDV	Median value of owner-occupied homes in $1000s

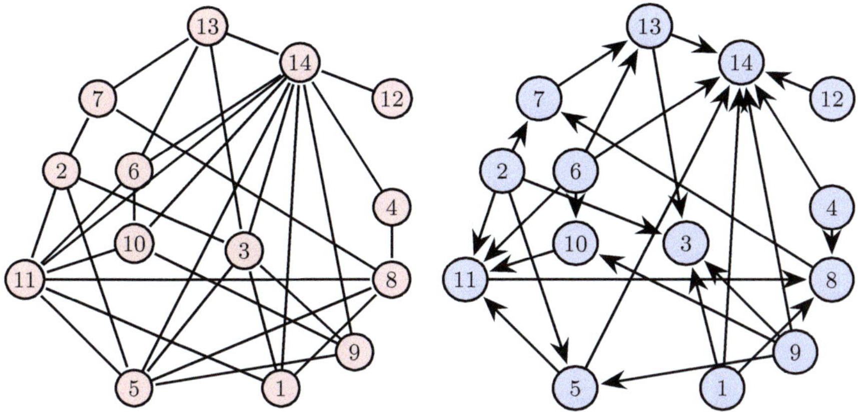

Fig. 5.4 Top: the skeleton consists of 32 undirected edges. After rule 0, there are 33 directed edge entries; after applying rules 1(a), 1(b), 1(c), and 1(d), the counts become 32, 30, 26, and 26, respectively. The bottom figure shows the resulting CPDAG

Problems 44–53

44. In Example 39, suppose we have a p-dimensional normal distribution, and let $\mathcal{X}, \mathcal{Y}, \mathcal{Z}$ be subsets of $\{1, \ldots, p\}$ of sizes s, t, u respectively, with $p = s+t+u$. If both $\mathcal{X} \perp\!\!\!\perp \mathcal{Z}$ and $\mathcal{Y} \perp\!\!\!\perp \mathcal{Z}$ hold, how many of the p^2 entries of the covariance matrix are zero?
45. Set the list gmG as follows:

```
p <- 10
n <- 1000
vars <-c("1", 2:p)  # if one element is a string, all are converted to strings
gGtrue <- randomDAG(p, prob = 0.5, V = vars)
gmG <- list(x = rmvDAG(n, gGtrue), g = gGtrue)
```

Using `pcalg`, output the true DAG, the estimated skeleton, and the estimated CPDAG.

46. Using the dataset gmD in `pcalg` (five nodes, taking 3,2,3,4,2 distinct values, respectively), one can perform the same procedure as in Fig. 5.1 by the following code:

```
data(gmD)
V <- colnames(gmD$x)
suffStat <- list(dm = gmD$x, nlev = c(3,2,3,4,2), adaptDF = FALSE)
pc.D <- pc(suffStat,
statistic indepTest = disCItest, alpha = 0.01, labels = V, verbose =
    TRUE)
par(mfrow = c(1,2))
plot(pc.D, main = "Estimated CPDAG")
plot(gmD$g, main = "True DAG")
```

Table 5.2 Conditional independences detected during skeleton learning on the Boston data using the PC algorithm ($m \geq 4$ detected none)

$m = 0$	$1 \perp\!\!\!\perp 4, 2 \perp\!\!\!\perp 4, 3 \perp\!\!\!\perp 4, 4 \perp\!\!\!\perp 5, 4 \perp\!\!\!\perp 6, 4 \perp\!\!\!\perp 7, 4 \perp\!\!\!\perp 9, 4 \perp\!\!\!\perp 10, 4 \perp\!\!\!\perp 12, 4 \perp\!\!\!\perp 13$
$m = 1$	$1 \perp\!\!\!\perp 6 \mid 5, 1 \perp\!\!\!\perp 7 \mid 5, 1 \perp\!\!\!\perp 7 \mid 8, 1 \perp\!\!\!\perp 12 \mid 9, 1 \perp\!\!\!\perp 12 \mid 10, 2 \perp\!\!\!\perp 12 \mid 1, 2 \perp\!\!\!\perp 12 \mid 3, 2 \perp\!\!\!\perp 14 \mid 3, 4 \perp\!\!\!\perp 11 \mid 14,$ $5 \perp\!\!\!\perp 12 \mid 1, 5 \perp\!\!\!\perp 12 \mid 3, 6 \perp\!\!\!\perp 9 \mid 2, 6 \perp\!\!\!\perp 9 \mid 3, 6 \perp\!\!\!\perp 9 \mid 5, 6 \perp\!\!\!\perp 12 \mid 2, 6 \perp\!\!\!\perp 12 \mid 3, 6 \perp\!\!\!\perp 12 \mid 5, 6 \perp\!\!\!\perp 12 \mid 10,$ $6 \perp\!\!\!\perp 12 \mid 11, 7 \perp\!\!\!\perp 9 \mid 5, 7 \perp\!\!\!\perp 9 \mid 8, 7 \perp\!\!\!\perp 10 \mid 5, 7 \perp\!\!\!\perp 10 \mid 8, 7 \perp\!\!\!\perp 11 \mid 5, 7 \perp\!\!\!\perp 12 \mid 3, 7 \perp\!\!\!\perp 12 \mid 5, 7 \perp\!\!\!\perp 12 \mid 8,$ $8 \perp\!\!\!\perp 12 \mid 5, 9 \perp\!\!\!\perp 12 \mid 1, 9 \perp\!\!\!\perp 12 \mid 10, 9 \perp\!\!\!\perp 13 \mid 10, 10 \perp\!\!\!\perp 12 \mid 1, 11 \perp\!\!\!\perp 12 \mid 1, 11 \perp\!\!\!\perp 12 \mid 14, 11 \perp\!\!\!\perp 13 \mid 14,$ $12 \perp\!\!\!\perp 13 \mid 14$
$m = 2$	$1 \perp\!\!\!\perp 5 \mid \{3, 9\}, 1 \perp\!\!\!\perp 5 \mid \{3, 10\}, 1 \perp\!\!\!\perp 5 \mid \{3, 11\}, 1 \perp\!\!\!\perp 5 \mid \{8, 9\}, 1 \perp\!\!\!\perp 5 \mid \{8, 10\}, 1 \perp\!\!\!\perp 5 \mid \{10, 11\},$ $1 \perp\!\!\!\perp 6 \mid \{8, 9\}, 1 \perp\!\!\!\perp 6 \mid \{8, 10\}, 1 \perp\!\!\!\perp 6 \mid \{8, 11\}, 1 \perp\!\!\!\perp 6 \mid \{9, 13\}, 1 \perp\!\!\!\perp 6 \mid \{10, 13\}, 1 \perp\!\!\!\perp 9 \mid \{2, 10\},$ $1 \perp\!\!\!\perp 9 \mid \{3, 10\}, 1 \perp\!\!\!\perp 9 \mid \{8, 10\}, 1 \perp\!\!\!\perp 13 \mid \{8, 14\}, 2 \perp\!\!\!\perp 6 \mid \{9, 13\}, 2 \perp\!\!\!\perp 8 \mid \{5, 9\}, 2 \perp\!\!\!\perp 9 \mid \{1, 3\},$ $2 \perp\!\!\!\perp 9 \mid \{3, 10\}, 2 \perp\!\!\!\perp 10 \mid \{1, 11\}, 2 \perp\!\!\!\perp 10 \mid \{5, 11\}, 2 \perp\!\!\!\perp 10 \mid \{7, 11\}, 3 \perp\!\!\!\perp 6 \mid \{5, 14\}, 3 \perp\!\!\!\perp 6 \mid \{9, 14\},$ $3 \perp\!\!\!\perp 7 \mid \{2, 5\}, 3 \perp\!\!\!\perp 7 \mid \{2, 8\}, 3 \perp\!\!\!\perp 7 \mid \{8, 14\}, 5 \perp\!\!\!\perp 13 \mid \{8, 14\}, 8 \perp\!\!\!\perp 9 \mid \{1, 7\}, 8 \perp\!\!\!\perp 9 \mid \{1, 10\},$ $8 \perp\!\!\!\perp 9 \mid \{1, 11\}, 8 \perp\!\!\!\perp 9 \mid \{1, 13\}, 8 \perp\!\!\!\perp 9 \mid \{3, 10\}, 8 \perp\!\!\!\perp 9 \mid \{4, 5\}, 8 \perp\!\!\!\perp 9 \mid \{4, 10\}, 8 \perp\!\!\!\perp 9 \mid \{5, 6\},$ $8 \perp\!\!\!\perp 9 \mid \{5, 10\}, 8 \perp\!\!\!\perp 9 \mid \{5, 11\}, 8 \perp\!\!\!\perp 9 \mid \{5, 13\}, 8 \perp\!\!\!\perp 9 \mid \{6, 10\}, 8 \perp\!\!\!\perp 9 \mid \{7, 10\}, 8 \perp\!\!\!\perp 9 \mid \{10, 13\},$ $9 \perp\!\!\!\perp 11 \mid \{10, 14\}, 10 \perp\!\!\!\perp 13 \mid \{3, 9\}, 10 \perp\!\!\!\perp 13 \mid \{3, 11\}, 10 \perp\!\!\!\perp 13 \mid \{5, 14\}, 10 \perp\!\!\!\perp 13 \mid \{8, 14\},$ $10 \perp\!\!\!\perp 13 \mid \{9, 14\}$
$m = 3$	$1 \perp\!\!\!\perp 2 \mid \{8, 11, 14\}, 3 \perp\!\!\!\perp 11 \mid \{1, 5, 14\}, 3 \perp\!\!\!\perp 11 \mid \{1, 8, 14\}, 3 \perp\!\!\!\perp 11 \mid \{2, 5, 14\}, 5 \perp\!\!\!\perp 10 \mid \{3, 7, 9\},$ $5 \perp\!\!\!\perp 10 \mid \{3, 8, 9\}, 6 \perp\!\!\!\perp 8 \mid \{10, 13, 14\}$

We would like to apply this method to the `asia` dataset to obtain a DAG. For this purpose, we convert "yes" and "no" into binary values 1 and 0:

```
library(bnlearn)
n <- dim(asia)[1]
p <- dim(asia)[2]
mat <- matrix(0, n, p)
for(i in 1:n)for(j in 1:p){
  if(asia[i,j]=="yes") mat[i,j] <- 1 else mat[i,j] <- 0
}
```

Fill in the blanks below to obtain a DAG from the `asia` dataset:

```
V <- colnames(asia) ## define sufficient statistics
suffStat <- list(dm = mat, nlev = #blank(1)#, adaptDF = TRUE)
pc.D <- pc(#blank(2)#, indepTest = disCItest, alpha = 0.05, labels = V
    )
plot(pc.D, main = "Estimated CPDAG")
```

47. In the function `K.merge`, why must the initial value of the matrix `K` not be set as the identity matrix?
48. In the function `skeleton`, why is the process carried out only when `length(S) >= m`? Also, under what condition does `s[i,j,k]` become `TRUE`?
49. Why must i and k not be adjacent in rule 0 and rule 1(a)?
50. For the 11 types of DAGs in Fig. 3.6, when X, Y, Z are relabeled as $1, 2, 3$, determine the values (`TRUE` or `FALSE`) of $S[i, j, k]$ for $(i, j, k) = (1, 2, 3), (1, 3, 2), (2, 1, 3), (2, 3, 1), (3, 1, 2), (3, 2, 1)$.
51. Consider the following undirected graph structure:

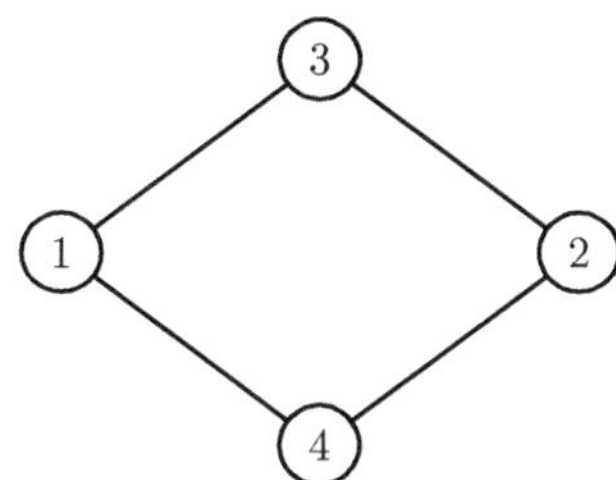

That is, there exist the following undirected edges among nodes 1, 2, 3, 4:

$$1 - 3, \quad 3 - 2, \quad 2 - 4, \quad 4 - 1$$

With $p = 4$, define $s[i, j, k]$ as

$$s[1, 2, 3] = s[2, 1, 3] = \texttt{FALSE}, \quad s[1, 2, 4] = s[2, 1, 4] = \texttt{TRUE}$$

All other $s[i, j, k]$ are `TRUE`

Apply `Post_PC(adj, s)` and show the resulting oriented graph. Also explain which rule (rule 0 or which case of rule 1) determined the orientation of each edge.

52. In Example 42, illustrate the CPDAG obtained immediately after applying rule 0 and rules 1(a)(b)(c)(d).
53. Write a function that takes as input vectors x, y of size n, a matrix z of size $n \times p$, and a significance level α and tests whether random variables X, Y are conditionally independent given $Z_1, \ldots, Z_p$. Then, generate two examples with $n = 100$, $p = 2$ by random sampling: one where $X \perp\!\!\!\perp Y \mid Z$ holds and one where it does not. Perform the test with your constructed function.

Chapter 6
LiNGAM

In this chapter, we describe in detail a method for identifying the causal order of variables, linear non-Gaussian acyclic model (LiNGAM) [23, 24]. Unlike the conventional Peter–Clark (PC) algorithm or score-based structure learning, LiNGAM assumes an additive noise model and directly identifies the causal order of variables.

First, we introduce direct LiNGAM, which assumes an ordering of variables in a linear structural model and detects upstream variables by computing residuals. Here, as theoretical background for the identifiability of causal order, we provide an intuitive explanation of the Darmois–Skitovitch theorem [5, 27]. Next, we present LiNGAM based on independent component analysis (ICA-LiNGAM) [23] and explain the flow of the algorithm together with the mathematical background of ICA.

Furthermore, as an extension of LiNGAM under the presence of confounders, we touch on the theoretical results of Wang and Drton and a structure recovery method based on shortest paths. These results show that non-Gaussianity broadens the possibilities for causal discovery.

(The programs in Chap. 6 assume the execution of the programs in Chaps. 4 and 5.)

6.1 Direct LiNGAM

Linear non-Gaussian model *(LiNGAM)* is a causal discovery method devised by Shohei Shimizu. There are several versions; in this chapter, we describe ICA-LiNGAM (2006) [23] and direct LiNGAM (2011) [24]. We begin with direct LiNGAM.

Suppose random variables X, Y can be expressed as

$$\begin{cases} X = e_1 \\ Y = aX + e_2 \end{cases} \tag{6.1}$$

J. Suzuki, *Graphical Models and Causal Discovery with R*,
https://doi.org/10.1007/978-981-95-4267-3_6

for some constant $a \in \mathbb{R}$ and mutually independent mean-zero random variables e_1, e_2, or as

$$\begin{cases} Y = e'_1 \\ X = a'Y + e'_2 \end{cases} \tag{6.2}$$

for some constant $a' \in \mathbb{R}$ and mutually independent mean-zero random variables e'_1, e'_2.

If the first model holds, we regard X as the cause and Y as the effect, and write $X \to Y$. If the second holds, we write $Y \to X$. This is called the *additive noise model* [11].

Model (6.1) is a statistical model in which X occurs independently of Y, and independent noise e_2 is added to a multiple of $X = e_1$. Conversely, (6.2) is a model in which Y occurs independently of X, and independent noise e'_2 is added to a multiple of $Y = e'_1$ (Fig. 6.1).

Here, the constants a, a' are determined to minimize mean squared error. That is, differentiating

$$\mathbb{E}_{XY}[\|Y - aX\|^2]$$

with respect to a and setting it to zero gives

$$-2\mathbb{E}_{XY}[X(Y - aX)] = 0, \tag{6.3}$$

so unless $X = 0$ almost surely,

$$a = \frac{\mathbb{E}_{XY}[XY]}{\mathbb{E}_X[X^2]}. \tag{6.4}$$

However, under (6.3) and (6.4), the covariance becomes

$$cov(e_1, e_2) = \mathbb{E}_{XY}[X(Y - aX)] = 0. \tag{6.5}$$

Similarly, $cov(e'_1, e'_2) = 0$ also holds. Note that when (X, Y) follows a normal distribution, $cov(X, Y) = 0$ implies $X \perp\!\!\!\perp Y$ (Proposition 1 in Chap. 2). In other

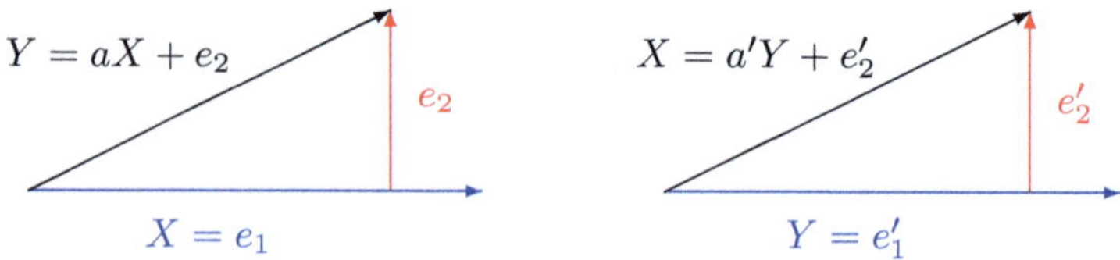

Fig. 6.1 In the statistical model (6.1), X occurs independently of Y, and independent noise e_2 is added to a multiple of $X = e_1$. Model (6.2) is the reverse

words, unless at least one pair of noises is nonindependent, both pairs become independent ($e_1 \perp\!\!\!\perp e_2$ and $e'_1 \perp\!\!\!\perp e'_2$), and identifiability is lost. Identifiability will be examined in detail in Sect. 6.3.

In practice, given independent samples from X, Y, we estimate which of (6.1) or (6.2) holds. Specifically, for n samples

$$x^n := (x_1, \ldots, x_n), \quad y^n := (y_1, \ldots, y_n) \in \mathbb{R}^n,$$

we subtract the sample means $\bar{x} := \frac{1}{n}\sum_{i=1}^{n} x_i$, $\bar{y} := \frac{1}{n}\sum_{i=1}^{n} y_i$ (center them), compute the empirical variances and covariance

$$v(x^n) := \frac{1}{n}\sum_{i=1}^{n} x_i^2, \quad v(y^n) := \frac{1}{n}\sum_{i=1}^{n} y_i^2, \quad c(x^n, y^n) := \frac{1}{n}\sum_{i=1}^{n} x_i y_i,$$

and then compute the residuals

$$y_x^n := y^n - x^n \cdot \frac{c(x^n, y^n)}{v(x^n)} \quad \text{and} \quad x_y^n := x^n - y^n \cdot \frac{c(x^n, y^n)}{v(y^n)}. \tag{6.6}$$

We then evaluate which of

$$x^n \perp\!\!\!\perp y_x^n \quad \text{or} \quad y^n \perp\!\!\!\perp x_y^n$$

is closer to independence.

One may also consider the mutual information

$$I(X, Y) := \int_{\mathcal{X}}\int_{\mathcal{Y}} f_{XY}(x, y) \log \frac{f_{XY}(x, y)}{f_X(x) f_Y(y)} dxdy,$$

where $\mathcal{X}, \mathcal{Y}$ are the ranges of X, Y. The fact that $I(X, Y) = 0$ if and only if $X \perp\!\!\!\perp Y$ (Sect. 3.4) provides a criterion. However, LiNGAM often uses HSIC as introduced in Chap. 4. Two approaches are possible:

1. Compare the values of the HSIC estimator (4.7) and select the smaller one.
2. Use the larger p-value obtained from HSIC.

HSIC values themselves are not comparable across different settings, and the first method lacks theoretical justification. Still, if independence holds, the HSIC estimate is close to zero and provides evidence. The second method takes more computation but yields more reliable results.

First, we implement method 1 in R. In the following function `LiNGAM.1(x,y)`, we evaluate the causal direction $X \to Y$. The smaller the returned value (HSIC), the more plausible $X \to Y$ is.

```
LiNGAM.1 <- function(x, y, proc = "HSIC") {
  # x, y: data vectors to be compared
  # proc: method to use ("HSIC" or "p.val")
  sigma2 <- 1  # variance parameter for RBF kernel
  # Define RBF (Radial Basis Function) kernel
  k.x <- function(x, y) exp(-norm(x - y, "2")^2 / (2 * sigma2))
  # Similarity measure for y uses the same kernel
  k.z <- k.x
  # Regression residual of y on x
  z <- y - sum(x * y) / sum(x^2) * x
  # Kernel matrices
  K.x <- K(k.x, x)
  K.z <- K(k.z, z)
  # HSIC (Hilbert-Schmidt Independence Criterion)
  result <- HSIC.2(K.x, K.z)
  if (proc == "HSIC") {
    return(result$statistics)
  } else {
    seq <- null.dist(result$eigen_values, 0.05)$z
    return(mean(result$statistics < seq))
  }
}
```

Example 43 Compare `LiNGAM.1(x,y)` and `LiNGAM.1(y,x)`. If the former is smaller, infer $X \to Y$; if the latter is smaller, infer $Y \to X$. Before running, define the following functions from Chap. 3: kernel matrix function `K`, centering function `tilde`, HSIC value function `HSIC.1`, and `null.dist` to set rejection regions. Then, using `LiNGAM.1`, compute the HSIC values and p-values for both directions:

```
> n <- 100
> x <- rnorm(n)
> y <- x + scale(rnorm(n)^2, center = TRUE)
> LiNGAM.1(x, y, proc="HSIC")
[1] 610.0936
> LiNGAM.1(y, x, proc="HSIC")
[1] 17058.65
> LiNGAM.1(x, y, proc="p.val")
[1] 0.3676
> LiNGAM.1(y, x, proc="p.val")
[1] 0
```

In this example, the true direction is $X \to Y$. Since `LiNGAM.1(x,y)` is smaller, the inference is correct. ■

6.2 Extension to Multiple Variables

Model (6.1) extends to the three-variable case X, Y, Z. If there exist constants $a, b, c \in \mathbb{R}$ and mutually independent random variables e_1, e_2, e_3 such that

$$\begin{cases} X = e_1 \\ Y = aX + e_2 \\ Z = bX + cY + e_3 \end{cases} \tag{6.7}$$

then we write $X \to Y \to Z$. The other five possible orders are treated analogously. First, set

$$a = \frac{\mathbb{E}_{XY}[XY]}{\mathbb{E}_X[X^2]}, \quad b = \frac{\mathbb{E}_{XZ}[XZ]}{\mathbb{E}_X[X^2]}, \quad c = \frac{\mathbb{E}_{YZ}[YZ]}{\mathbb{E}_Y[Y^2]},$$

$$a' = \frac{\mathbb{E}_{XY}[XY]}{\mathbb{E}_Y[Y^2]}, \quad b' = \frac{\mathbb{E}_{XZ}[XZ]}{\mathbb{E}_Z[Z^2]}, \quad c' = \frac{\mathbb{E}_{YZ}[YZ]}{\mathbb{E}_Z[Z^2]},$$

and compare

$$X \perp\!\!\!\perp \{Y - aX,\ Z - bX\},\ Y \perp\!\!\!\perp \left\{Z - cY,\ X - a'Y\right\},\ Z \perp\!\!\!\perp \left\{X - b'Z,\ Y - c'Z\right\}.$$

That is, we select the most upstream variable and then test its independence from the two residualized variables (i.e., variables after removing its effect). For example, if the first condition is judged to be true (i.e., X is most upstream), then we compare

$$(Y - aX) \perp\!\!\!\perp \left((Z - bX) - (Y - aX) \cdot \frac{\mathbb{E}_{XYZ}[(Z - bX)(Y - aX)]}{\mathbb{E}_{XY}[(Y - aX)^2]}\right),$$

$$(Z - bX) \perp\!\!\!\perp \left((Y - aX) - (Z - bX) \cdot \frac{\mathbb{E}_{XYZ}[(Y - aX)(Z - bX)]}{\mathbb{E}_{XZ}[(Z - bX)^2]}\right).$$

If the former holds, we conclude $X \to Y \to Z$; if the latter holds, we conclude $X \to Z \to Y$.

More concretely, for observed data

$$x^n = (x_1, \ldots, x_n), \qquad y^n = (y_1, \ldots, y_n), \qquad z^n = (z_1, \ldots, z_n),$$

we first estimate which of

$$x^n \perp\!\!\!\perp \{y^n_x, z^n_x\} \quad \text{or} \quad y^n \perp\!\!\!\perp \{z^n_y, x^n_y\} \quad \text{or} \quad z^n \perp\!\!\!\perp \{x^n_z, y^n_z\}$$

is true. Here, x^n_z, y^n_z, z^n_x, z^n_y are defined analogously to (6.6). If, for instance, $x^n \perp\!\!\!\perp \{y^n_x, z^n_x\}$ holds, define the residuals

$$y^n_{xz} := y^n_x - \frac{c(y^n_x, z^n_x)}{v(z^n_x)} z^n_x, \qquad z^n_{xy} := z^n_x - \frac{c(y^n_x, z^n_x)}{v(y^n_x)} y^n_x, \tag{6.8}$$

and then estimate which of

$$y^n_x \perp\!\!\!\perp z^n_{xy} \quad \text{or} \quad z^n_x \perp\!\!\!\perp y^n_{xz}$$

is true. If the former holds, we obtain $X \to Y \to Z$. Overall, the decision procedure is as follows:

$$\begin{cases} x^n \perp\!\!\!\perp \{y_x^n, z_x^n\} \Longrightarrow \begin{cases} y_x^n \perp\!\!\!\perp z_{xy}^n \Longrightarrow X \to Y \to Z \\ z_x^n \perp\!\!\!\perp y_{xz}^n \Longrightarrow X \to Z \to Y \end{cases} \\ y^n \perp\!\!\!\perp \{z_y^n, x_y^n\} \Longrightarrow \begin{cases} z_y^n \perp\!\!\!\perp x_{yz}^n \Longrightarrow Y \to Z \to X \\ x_y^n \perp\!\!\!\perp z_{yx}^n \Longrightarrow X \to Z \to Y \end{cases} \\ z^n \perp\!\!\!\perp \{x_z^n, y_z^n\} \Longrightarrow \begin{cases} x_z^n \perp\!\!\!\perp y_{zx}^n \Longrightarrow Z \to X \to Y \\ y_z^n \perp\!\!\!\perp x_{zy}^n \Longrightarrow Z \to Y \to X \end{cases} \end{cases}$$

Moreover,

$$x_{yz}^n = x_{zy}^n, \qquad y_{zx}^n = y_{xz}^n, \qquad z_{xy}^n = z_{yx}^n$$

hold (Fig. 6.2). Indeed, z_{xy}^n can be written in terms of x^n, y^n, z^n as

$$z^n - \frac{v(y^n)c(x^n, z^n) - c(x^n, y^n)c(y^n, z^n)}{v(x^n)v(y^n) - c(x^n, y^n)^2} x^n - \frac{v(x^n)c(y^n, z^n) - c(x^n, y^n)c(x^n, z^n)}{v(x^n)v(y^n) - c(x^n, y^n)^2} y^n, \tag{6.9}$$

(see the appendix at the end of the chapter for a proof). Swapping x^n and y^n in (6.9) only exchanges the second and third terms, leaving the overall value unchanged. We

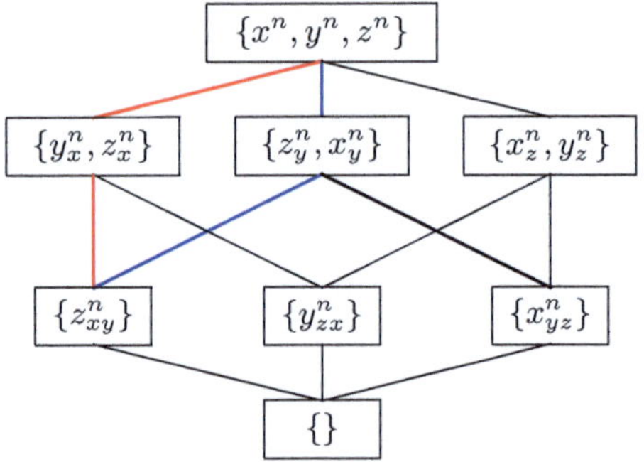

Fig. 6.2 Residuals of upstream variables propagate downstream. The paths $X \to Y \to Z$ (shown in red) and $Y \to X \to Z$ (shown in blue) merge at $\{z_{xy}^n\}$, but even though the order of X and Y differs, the residuals are equal: $z_{xy} = z_{yx}$

$$\begin{aligned}
X \to Y \to Z &: \{x^n, y^n, z^n\} \to \{y_x^n, z_x^n\} \to \{z_{xy}^n\} \to \{\} \\
Y \to X \to Z &: \{x^n, y^n, z^n\} \to \{z_y^n, x_y^n\} \to \{z_{yx}^n\} \to \{\} \\
Y \to Z \to X &: \{x^n, y^n, z^n\} \to \{z_y^n, x_y^n\} \to \{x_{yz}^n\} \to \{\} \\
X \to Z \to Y &: \{x^n, y^n, z^n\} \to \{y_x^n, z_x^n\} \to \{y_{xz}^n\} \to \{\} \\
Z \to X \to Y &: \{x^n, y^n, z^n\} \to \{y_z^n, x_z^n\} \to \{y_{zx}^n\} \to \{\} \\
Z \to Y \to X &: \{x^n, y^n, z^n\} \to \{x_z^n, y_z^n\} \to \{x_{zy}^n\} \to \{\}
\end{aligned}$$

omit the proof, but even when the number of variables is $p \geq 4$, the value of the residual does not depend on the order of the subscripts (e.g., x^n_{yz} is independent of the order of y, z).

In the multivariate case as well, we determine a variable order so that the residuals become independent. We first extend the LiNGAM routine so it applies to three or more variables.

```
# Before running, define the following Chapter 3 functions:
# K: computes a kernel matrix
# tilde: centers a kernel matrix
# HSIC.2: computes the HSIC statistic
# null.dist: sets a rejection region from eigenvalues and a chosen alpha

# LiNGAM computation (choose HSIC or p-value)
LiNGAM.2 <- function(i, z, proc="HSIC") {
  sigma2 <- 1  # kernel scale parameter
  # Gaussian kernel for similarities
  k.x <- function(x, y) exp(-norm(x - y, "2")^2 / (2 * sigma2))
  # sample size
  n <- length(z[[i]])
  # number of variables
  m <- length(z)
  # kernel matrix for the i-th variable
  K.x <- K(k.x, z[[i]])
  # initialize K.z as identity (elementwise multiplicative accumulator)
  K.z <- matrix(1, n, n)
  # process all other variables
  index <- setdiff(1:m, i)
  for (j in index) {
    # residualize z[[j]] on z[[i]]
    z[[j]] <- z[[j]] - sum(z[[i]] * z[[j]]) / sum(z[[i]] ** 2) * z[[i]]
    # accumulate kernel matrices elementwise
    K.z <- K.z * K(k.x, z[[j]])
  }
  # compute HSIC statistic
  result <- HSIC.2(K.x, K.z)
  if (proc == "HSIC") {
    HSIC <- result$statistics
    return(list(HSIC=HSIC, Z=z[-i]))
  } else if (proc == "p.val"){
    seq <- null.dist(result$eigen_values, 0.05)$z
    p.val <- mean(result$statistics < seq)
    return(list(p.val=p.val, Z=z[-i]))
  }
}
```

Using `LiNGAM.2`, we define the function `search.HSIC` that searches for a variable ordering based on HSIC estimates. Given Z, a list whose elements are the data vectors of each variable, `search.HSIC` searches for an order by evaluating independence via HSIC.

Initially, call `search.HSIC` without specifying `index`; then `index` is set to `1:m` (where `m` is the length of `Z`). From these `m` candidates, we find the most

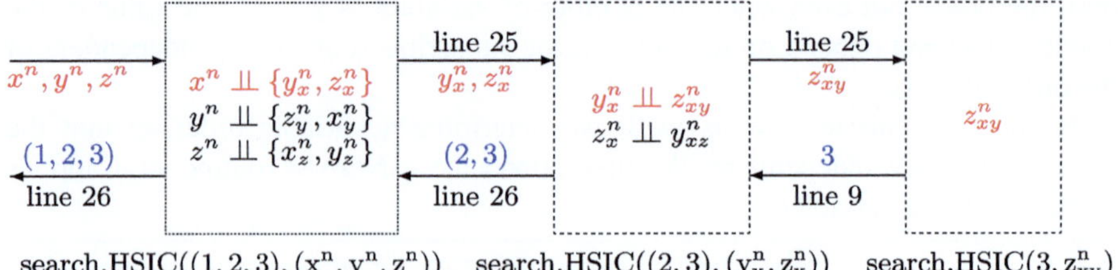

Fig. 6.3 When computing search.HSIC((1, 2, 3), (x^n, y^n, z^n)), if variable 1 is judged to be the most upstream, at line 25, it requests the computation of search.HSIC((2, 3), (y^n_x, z^n_x)). In the next process to the right, if variable 2 is judged to be the next upstream, then at line 25, it requests the computation of search.HSIC(3, z^n_{xy}). However, in the process on the right, since there is nothing further to compare, it returns the value 3 from line 9. In the process to the left, it returns the value (2,3) from line 26. Finally, the leftmost process returns the value (1,2,3) as the final output

upstream variable. If that index is `k`, we store its residualized set in `W` and call `search.HSIC(W, setdiff(index,k))`. The resulting order of the remaining variables is stored in `index.3`, and we output the final order by placing `k` at the front.

Except for the very first call, `search.HSIC` is called recursively from within itself, passing a subset of `1:m` as `index`. Such a function calling itself is a *recursive call*. Finally, `search.HSIC` returns a variable order that satisfies the LiNGAM criterion described above (Fig. 6.3).

```
1  search.HSIC <- function(Z, index=NULL) {
2    m <- length(Z)
3    # if index is omitted, treat as top-level call
4    if (length(index) == 0) index <- 1:m
5    # if fewer than two variables remain, return index
6    if (m < 2) return(index)
7    # initialize the current minimum HSIC
8    min <- Inf
9    # loop over candidates
10   for (i in 1:m) {
11     # compute HSIC via LiNGAM.2
12     result <- LiNGAM.2(i, Z, proc="HSIC")
13     # update best HSIC and record residuals and index
14     if (result$HSIC < min) {
15       min <- result$HSIC
16       W <- result$Z    # store residualized set
17       k <- index[i]
18     }
19   }
20   # recurse on the residualized set, then prepend k
21   index.3 <- search.HSIC(W, setdiff(index, k))
22   return(c(k, index.3))
23 }
```

Similarly, using `LiNGAM.2`, one can define `search.p.val`, which searches for an order based on p-values (Problem 55).

Example 44 We conducted an experiment to infer causal direction on the following synthetic data. Although a single trial does not suffice for confirmation, `search.p.val` often requires more computation but tends to yield correct answers more frequently.

```
> n <- 100
> x <- scale(rnorm(n)^2, center  = TRUE)
> y <- 2*x + scale(rnorm(n)^2, center  = TRUE)
> z <- -3*x + 4*y + scale(rnorm(n)^2, center  = TRUE)
> X <- list(x,y,z)
> search.HSIC(X)
[1] 1 2 3
> search.p.val(X)
[1] 1 2 3
```

In this experiment, the outputs of `search.HSIC` and `search.p.val` coincide. However, there are cases where the latter is correct while the former is not. ■

Finally, let us apply the LiNGAM ordering search to real data.

Example 45 We applied `search.HSIC` to the Boston dataset. The following order was obtained:

```
> library(MASS)
> Boston.101_200 <- as.data.frame(scale(Boston[101:200, ],
  center = TRUE, scale = FALSE))
> search.HSIC(as.list(Boston.101_200))
[1] 5 4 6 1 8 9 11 13 14 3 2 7 10 12
```

This result may disagree with the order produced by the PC algorithm in Chap. 5. Indeed, the PC algorithm constructs a DAG using only conditional independence information, whereas LiNGAM introduces the additive noise model as an alternative criterion and imposes an order on variables that would otherwise remain unordered. ■

If we follow the same procedure as in Example 45 but run `search.p.val(Boston)`, all p-values become zero and no order is obtained. In other words, LiNGAM requires the assumption that the noises are independent–equivalently, in (6.1) and (6.2), either $e_1 \perp\!\!\!\perp e_2$ or $e'_1 \perp\!\!\!\perp e'_2$ holds. In real data, however, this assumption often fails (confounding is common in practice). We address this issue in Sect. 5.4.

6.3 Identifiability

So far, we have described concrete procedures in LiNGAM. We now discuss theoretical properties.

A basic question is, if both (6.1) and (6.2) hold simultaneously, which direction, $X \to Y$ or $Y \to X$, should we assert? LiNGAM assumes that at most one of (6.1) and (6.2) holds. When exactly one holds, the causal direction is said to be **identifiable**; if both hold, it is not identifiable.

Example 46 Suppose in (6.1) that $a = 1$, $e_1, e_2 \sim N(0, 1)$, and $e_1 \perp\!\!\!\perp e_2$. Then in (6.2), for any $a' \in \mathbb{R}$, we can set

$$\begin{cases} e_1' = Y = e_1 + e_2 \\ e_2' = X - a'Y = e_1 - a'(e_1 + e_2) = (1 - a')e_1 - a'e_2 \end{cases}$$

The covariance of e_1', e_2' is $1 - 2a'$, so choosing $a' = 1/2$ makes this covariance zero. Since e_1, e_2 are normal, their linear combinations e_1', e_2' are also normal. By stability (Propositions 3) and 1, we have $e_1' \perp\!\!\!\perp e_2'$. Hence, both (6.1) and (6.2) can hold simultaneously. ■

We next seek general conditions under which $e_1 \perp\!\!\!\perp e_2$ and $e_1' \perp\!\!\!\perp e_2'$ cannot both hold. Assuming both representations are possible,

$$\begin{cases} X = e_1 \\ Y = aX + e_2 \end{cases} \begin{cases} Y = e_1' \\ X = a'Y + e_2' \end{cases} \Longrightarrow \begin{cases} e_1' = ae_1 + e_2 \\ e_1 = a'e_1' + e_2' \end{cases}$$

and since

$$e_2' = e_1 - a'e_1' = e_1 - a'(ae_1 + e_2) = (1 - aa')e_1 - a'e_2,$$

we obtain

$$\begin{cases} e_1' = ae_1 + e_2 \\ e_2' = (1 - aa')e_1 - a'e_2 \end{cases} \tag{6.10}$$

When $e_1 \perp\!\!\!\perp e_2$, letting the variances of e_1, e_2 be σ_1^2, σ_2^2, the covariance of e_1', e_2' is $a(1 - aa')\sigma_1^2 - a'\sigma_2^2$. For any $a \in \mathbb{R}$, choosing

$$a' = \frac{a\sigma_1^2}{a^2\sigma_1^2 + \sigma_2^2} \tag{6.11}$$

makes this covariance zero. Thus if e_1, e_2 are normal, then for any a, there exists a' such that both $e_1 \perp\!\!\!\perp e_2$ and $e_1' \perp\!\!\!\perp e_2'$ hold.

The converse also holds. Using the following fact, we show that both $e_1 \perp\!\!\!\perp e_2$ and $e_1' \perp\!\!\!\perp e_2'$ can hold only when all these variables are normally distributed.

Proposition 19 (Darmois–Skitovitch, 1953) *Let $X_1, \ldots, X_m$ be mutually independent, nondegenerate random variables, and let Y_1, Y_2 be independent of each other and related by constants $\alpha_1, \ldots, \alpha_m, \beta_1, \ldots, \beta_m \in \mathbb{R}$ as*

$$Y_1 = \alpha_1 X_1 + \cdots + \alpha_m X_m, \tag{6.12}$$

$$Y_2 = \beta_1 X_1 + \cdots + \beta_m X_m. \tag{6.13}$$

Then for every $i = 1, \ldots, m$ with $\alpha_i \beta_i \neq 0$, X_i is normally distributed. Assume there do not exist $i \neq j$ with $\alpha_i \beta_j - \alpha_j \beta_i = 0$.[1]

Proof See the appendix at the end of the chapter.

Apply Proposition 19 with $Y_1 = e'_1$, $Y_2 = e'_2$ and $m = 2$ in (6.10). Since (6.11) is necessary for $e_1 \perp\!\!\!\perp e_2$ and $e'_1 \perp\!\!\!\perp e'_2$ to hold, we can set

$$(\alpha_1, \alpha_2, \beta_1, \beta_2) = (a, 1, 1 - aa', -a') = \left(a, 1, \frac{\sigma_2^2}{a^2\sigma_1^2 + \sigma_2^2}, -\frac{a\sigma_1^2}{a^2\sigma_1^2 + \sigma_2^2}\right).$$

Hence,

$$\alpha_1 \beta_1 = \frac{a\sigma_2^2}{a^2\sigma_1^2 + \sigma_2^2}, \qquad \alpha_2 \beta_2 = -\frac{a\sigma_1^2}{a^2\sigma_1^2 + \sigma_2^2}.$$

By Proposition 19, except for the trivial case $a = 0$ (equivalently $a' = 0$ and $X \perp\!\!\!\perp Y$, Problem 57), both e_1 and e_2 must be normal. The discussion so far can be summarized as follows.

Proposition 20 (Shimizu et al. [24]) *Assume $X \not\perp\!\!\!\perp Y$. Then the following are equivalent: non-identifiability (i.e., both $e_1 \perp\!\!\!\perp e_2$ and $e'_1 \perp\!\!\!\perp e'_2$ hold), e_1, e_2 are normal, and e'_1, e'_2 are normal.*

We next consider extending Proposition 20 to the multivariate case.

For multiple variables, we give a sufficient condition for identifiability.

Proposition 21 *If at least $p - 1$ of the noises $e_1, \ldots, e_p$ are non-Gaussian, then the order of $X_1, \ldots, X_p$ is identifiable.*

Proof: See the appendix at the end of the chapter.

Proposition 21 is only a sufficient condition. Indeed, there exist identifiable cases even when two or more noises are Gaussian (Problems 59 and 60).

6.4 Independent Component Analysis (ICA)

Before discussing ICA LiNGAM, we first explain *independent component analysis (ICA)*.

[1] Otherwise, $\alpha_i X_i + \alpha_j X_j$ becomes a constant multiple of $\beta_i X_i + \beta_j X_j$, so $\alpha_i X_i + \alpha_j X_j$ can be treated as a single random variable.

In direct LiNGAM, we mainly assumed a *structural equation model* (SEM). Thus, for $i = 1, \dots, p$,

$$X_i = \sum_{i=1}^{i-1} b_{i,j} X_j + e_i,$$

where $e_1, \dots, e_p$ are mutually independent. Let $B = (b_{i,j})$ have nonzero elements only strictly below the diagonal, and set $X = [X_1, \dots, X_p]^\top$, $e = [e_1, \dots, e_p]^\top$. Then we can write

$$\begin{bmatrix} X_1 \\ \vdots \\ X_p \end{bmatrix} = B \begin{bmatrix} X_1 \\ \vdots \\ X_p \end{bmatrix} + \begin{bmatrix} e_1 \\ \vdots \\ e_p \end{bmatrix} \quad \text{or} \quad \begin{bmatrix} X_1 \\ \vdots \\ X_p \end{bmatrix} = (I - B)^{-1} \begin{bmatrix} e_1 \\ \vdots \\ e_p \end{bmatrix}. \tag{6.14}$$

Here, direct LiNGAM treats B as unknown and, from n realizations of $X \in \mathbb{R}^p$, estimates the mutually independent $e_1, \dots, e_p$. In general, given a random vector $X \in \mathbb{R}^p$, the task of estimating at most p independent random variables $S = [s_1, \dots, s_m]^\top$ such that $X = WS$ is called independent component analysis.

Without loss of generality below, assume X and e both have mean zero and that at most one of the m components s_j is Gaussian. Indeed, if $X \sim N(0, \sigma_1^2)$ and $Y \sim N(0, \sigma_2^2)$, then by stability (Proposition 3) , $X + Y \sim N(0, \sigma_1^2 + \sigma_2^2)$, and (proof omitted) there exist independent $\tilde{X}, \tilde{Y}$ with $X' + Y' \sim N(0, \sigma_1^2 + \sigma_2^2)$ that necessarily follow mean-zero Gaussian distributions (Cramér' s theorem [17]). In that case, again by stability, $\tilde{X} \sim N(0, \tilde{\sigma}_1^2)$, $\tilde{Y} \sim N(0, \tilde{\sigma}_2^2)$ with $\tilde{\sigma}_1^2 > 0, \tilde{\sigma}_2^2 > 0$, and $\tilde{\sigma}_1^2 + \tilde{\sigma}_2^2 = \sigma_1^2 + \sigma_2^2$ can take any values satisfying this sum constraint. Thus, one cannot separate Gaussian components from a linear combination of mean-zero Gaussian random variables.

Therefore, extracting independent components amounts to extracting non-Gaussian components. There are several ways to implement ICA, but a common approach repeatedly extracts the component that deviates most from normality, then searches–within the subspace orthogonal to it–for the next component with the largest deviation from normality.

In the CRAN package `fastICA`, one first centers and scales the data $X \in \mathbb{R}^{n \times p}$ and then whitens it. Let $V := \frac{1}{n} X^\top X$ be the sample covariance matrix, $U = [u_1, \dots, u_p] \in \mathbb{R}^{p \times p}$ the matrix of eigenvectors, and $D = \text{diag}(\lambda_1, \dots, \lambda_p)$, $D^{-1/2} = \text{diag}(\lambda_1^{-1/2}, \dots, \lambda_p^{-1/2})$ the diagonal matrices of eigenvalues and their inverse square roots. Defining $\tilde{X} := XUD^{-1/2} \in \mathbb{R}^{n \times p}$, we have $U^\top U = I \in \mathbb{R}^{p \times p}$, and from $VU = UD$,

$$\begin{aligned} \frac{1}{n} \tilde{X}^\top \tilde{X} &= \frac{1}{n} (XUD^{-1/2})^\top XUD^{-1/2} = D^{-1/2} U^\top \cdot \frac{1}{n} X^\top X \cdot UD^{-1/2} \\ &= D^{-1/2} U^\top VUD^{-1/2} = D^{-1/2} U^\top UDD^{-1/2} = I, \end{aligned}$$

so the columns of $\tilde{X}$ are uncorrelated. Below, we relabel such an $\tilde{X}$ simply as $X \in \mathbb{R}^{n \times p}$.

For $X = (X_{i,j})$, define $s_i = \sum_{j=1}^{p} X_{i,j} w_j$ for $i = 1, \ldots, n$, and under $\sum_{j=1}^{p} w_j^2 = 1$, consider finding $w = (w_1, \ldots, w_p)$ that maximizes $(1/n) \sum_{i=1}^{n} G(s_i)$. Here, $G(s)$ measures the degree of non-Gaussianity of s, and several candidates exist (Problem 61).

For probability density functions f, g, define the Kullback–Leibler divergence

$$D(f \| g) := \mathbb{E}_X\left[\log \frac{f(X)}{g(X)}\right] = \int_{-\infty}^{\infty} f(x) \log \frac{f(x)}{g(x)}\, dx.$$

This quantity is nonnegative and is zero if and only if $\log\bigl(f(X)/g(X)\bigr) = 0$ with probability one, i.e., $f(X) = g(X)$. Let

$$f_{gauss}(x) = \frac{1}{\sqrt{2\pi\sigma^2}} \exp\left(-\frac{x^2}{2\sigma^2}\right),$$

and assume

$$\int_{-\infty}^{\infty} x^2 f(x)\, dx = \int_{-\infty}^{\infty} x^2 f_{gauss}(x)\, dx = \sigma^2. \tag{6.15}$$

Then for any density f,

$$h(f) := \int_{-\infty}^{\infty} -f(x) \log f(x)\, dx \le h(f_{gauss}) = \frac{1}{2} \log(2\pi\sigma^2 e) \tag{6.16}$$

holds (Problem 62).

In general, one maximizes the *negentropy* $h(f_{gauss}) - h(f)$, i.e.,

$$G(s) = -\log f(s).$$

However, this formulation requires estimating the density f, which is not straightforward.

For computational efficiency, `fastICA` uses $G(s) = \log\cosh(s)$ or $G(s) = -e^{-s^2}$. In practice, one solves the equation $G'(s) = g(s) = 0$, applying Newton–Raphson. See the original paper for details [9].

For verification, we ran ICA using CRAN's `fastICA`.

```r
# Load required library
library(fastICA)
# Generate independent signals S (5000x2 matrix)
S <- matrix(runif(10000), 5000, 2)
# Define a mixing matrix A
A <- matrix(c(1, 1, -1, 3), 2, 2, byrow = TRUE)
# Create observed data X by mixing S with A
```

```
X <- S %*% A
# Apply ICA
a <- fastICA(X, 2,
                alg.typ = "parallel",
                fun = "logcosh",
                alpha = 1,
                method = "C",
                row.norm = FALSE,
                maxit = 200,
                tol = 0.0001,
                verbose = TRUE)
# Show three plots side by side
par(mfrow = c(1, 3))
plot(a$X, main = "Preprocessed Data", col = "blue", pch = 20)
plot(a$X %*% a$K, main = "PCA Result", col = "red", pch = 20)
plot(a$S, main = "ICA Result", col = "green", pch = 20)
```

The output is shown in Fig. 6.4.

We also performed signal recovery from mixed signals via ICA (Problem 63). We first generated independent signals–a sine wave and a linearly varying signal–and formed observed data (mixed signals) by linearly combining them with a mixing matrix. We then used `fastICA` to estimate independent components from the mixed signals and finally visualized the original signals, the mixed signals, and the ICA-separated signals (Fig. 6.5).

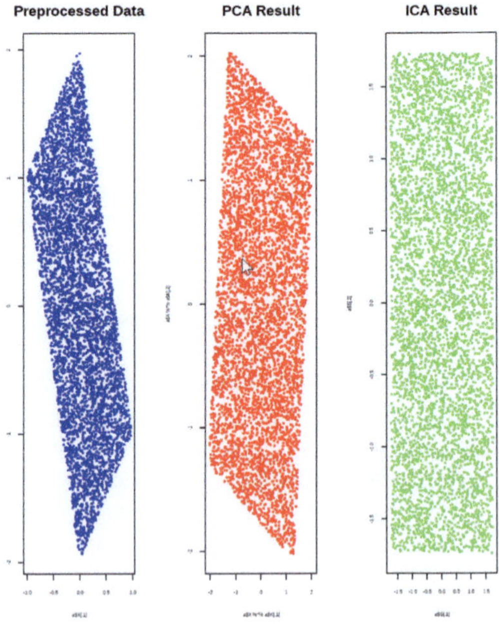

Fig. 6.4 ICA was run and the process visualized: centering/scaling, whitening, and ICA results are displayed

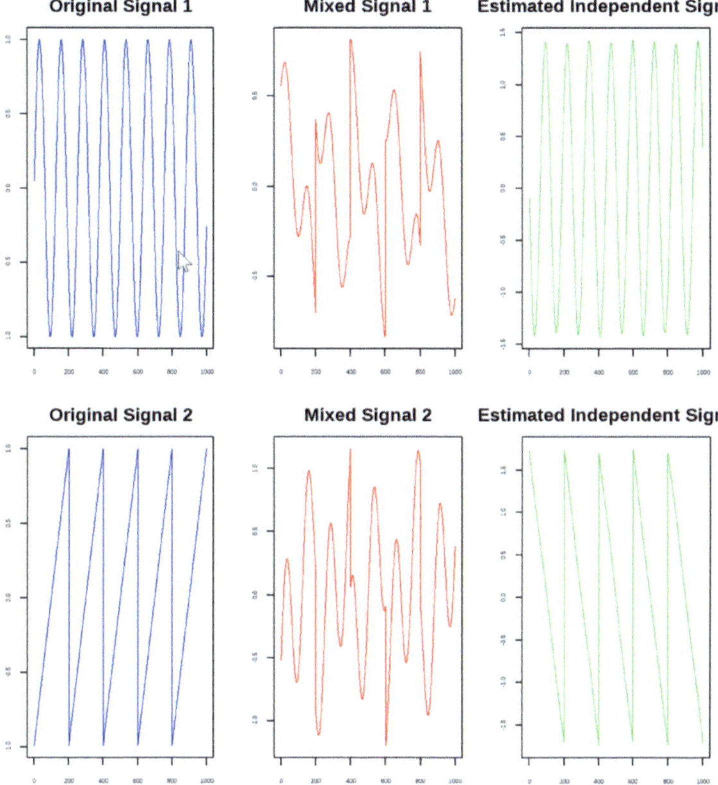

Fig. 6.5 Separations of composed waves

6.5 ICA LiNGAM

ICA LiNGAM is an algorithm that determines a variable ordering by extracting independent components using ICA.

In `fastICA`, the learned `W` is the weight matrix for the whitened data $\tilde{X} = XK$ with $K = UD^{-1/2}$ applied to X. That is,

$$S = \tilde{X}W = (XK)W = X \cdot KW.$$

Below, for simplicity of notation, we write KW as W.

Although ICA does not care about the order of extracted components, from (6.14), we expect W to be $I - B$ (with I the identity and B a strictly lower triangular matrix with zero diagonal). We regard the vector e as the independent components. Thus, we seek a permutation of the rows of W so that it becomes a lower triangular matrix with all diagonal entries equal to 1.

Let P be a square matrix obtained from the $p \times p$ identity by swapping the i-th and j-th diagonal 1s to the off-diagonal (i, j) and (j, i) positions (and zeros elsewhere); then left-multiplying a matrix A by P swaps the i-th and j-th rows of A. More generally, if P has exactly one 1 in each row and column (and zeros otherwise), then left-multiplying by P permutes the rows of A. Such a matrix is called a *permutation matrix*. If the i-th row of P has its 1 in column $\pi(i)$, then specifying P is equivalent to specifying a bijection

$$\pi : (1, \ldots, p) \mapsto (\pi(1), \ldots, \pi(p)),$$

of which there are $p!$ possibilities. Hence, choosing a permutation matrix P is equivalent to choosing an order of the rows of A (Problem 65).

By left-multiplying W with a permutation matrix P, we can reorder the rows of W. If the estimation is correct, the diagonal entries of W will be nonzero. ICA LiNGAM chooses P that minimizes the sum of reciprocals of the absolute values of the diagonal entries of PW.

Next, let $PW = (c_{ij})$ and scale it so that its diagonal becomes 1. To this end, form the diagonal matrix $D = \mathrm{diag}(c_{11}, \ldots, c_{pp})$ from the diagonal entries of PW and left-multiply by D^{-1}:

$$D^{-1}PW = \begin{bmatrix} c_{11}^{-1} & 0 & \cdots & 0 \\ 0 & c_{22}^{-1} & \cdots & 0 \\ \vdots & \vdots & \ddots & \vdots \\ 0 & 0 & \cdots & c_{pp}^{-1} \end{bmatrix} \begin{bmatrix} c_{11} & c_{12} & \cdots & c_{1p} \\ c_{21} & c_{22} & \cdots & c_{2p} \\ \vdots & \vdots & \ddots & \vdots \\ c_{p1} & c_{p2} & \cdots & c_{pp} \end{bmatrix} = \begin{bmatrix} 1 & \frac{c_{12}}{c_{11}} & \cdots & \frac{c_{1p}}{c_{11}} \\ \frac{c_{21}}{c_{22}} & 1 & \cdots & \frac{c_{2p}}{c_{22}} \\ \vdots & \vdots & \ddots & \vdots \\ \frac{c_{p1}}{c_{pp}} & \frac{c_{p2}}{c_{pp}} & \cdots & 1 \end{bmatrix}.$$

Here, we assume that the upper-triangular part is close to zero. Subtracting 1 from the diagonal gives an estimate of $W = I - B$.

Since there are $p!$ possible permutations P, one would, in principle, compare the sums of reciprocals of the absolute diagonal entries of PW over all permutations. However, this becomes intractable when $p > 10$. Therefore, we reduce the problem to finding a mapping π that minimizes $\sum_{i=1}^{p} |W_{i,\pi(i)}|^{-1}$. The CRAN package `clue` provides an assignment problem solver (using the Hungarian method) to obtain an exact solution.

Finally, we run `LINGAM` from CRAN's `pcalg`.

Example 47 (Estimating Causal Structure with LiNGAM (Two Variables)) We illustrate a simple example of estimating causal structure using linear non-Gaussian acyclic model (LiNGAM). The following program generates data with two variables, X_1 and X_2, where the true model has the causal relation $X_2 \to X_1$. We then apply LiNGAM and examine how well the estimated causal structure matches the true structure.

```
library(pcalg)
set.seed(1234) # set seed for reproducibility
n <- 500       # sample size
# Generate noises (exogenous variables)
eps1 <- sign(rnorm(n)) * sqrt(abs(rnorm(n))) # noise 1: signed sqrt-transformed normal
eps2 <- runif(n) - 0.5                        # noise 2: Uniform(-0.5, 0.5)
# Generate x2
x2 <- 3 + eps2
# Generate x1
x1 <- 0.9*x2 + 7 + eps1
# True DAG: x2 → x1
trueDAG <- cbind(c(0,1),c(0,0))
# Data matrix
X <- cbind(x1,x2)
# Apply LiNGAM
res <- lingam(X)
# Show true DAG
cat("【True】 DAG:\n")
show(trueDAG)
# Show estimated DAG
cat("【Estimated】 DAG:\n")
show(as(res, "amat"))
# Show true intercepts
cat("\【nTrue】 intercepts:\n")
show(c(7,3))
# Show estimated intercepts
cat("【Estimated】 intercepts:\n")
show(res$ci)
# Show true noise standard deviations (sample SD)
cat("\【nTrue noise】 SD:\n")
show(c(sd(eps1), sd(eps2)))
# Show estimated noise standard deviations
cat("【Estimated noise】 SD:\n")
show(res$stde)
```

Readers are encouraged to run the program themselves (Problem 64).

The causal structure estimated by LiNGAM matches the true DAG. The estimated intercepts and noise standard deviations are also close to their true values. This demonstrates that, under the non-Gaussianity assumption, LiNGAM can identify the correct causal structure. ■

Example 48 (Estimating Causal Structure with LiNGAM (Four Variables)) We estimate a causal structure with four variables. Taking x_2 as the source, x_1 and x_3 are generated under the influence of x_2, and x_4 is influenced by both x_1 and x_3. Thus, the true causal structure is

$$x_2 \to x_1, \quad x_2 \to x_3, \quad x_1 \to x_4, \quad x_3 \to x_4.$$

We apply LiNGAM to these data and compare the estimated structure with the true one.

```
library(pcalg)
set.seed(123)  # reproducibility
n <- 500       # sample size
# Generate noises (exogenous variables)
eps1 <- sign(rnorm(n)) * sqrt(abs(rnorm(n)))   # signed sqrt-transformed normal
eps2 <- runif(n) - 0.5                          # Uniform(-0.5, 0.5)
eps3 <- sign(rnorm(n)) * abs(rnorm(n))^(1/3)   # signed 1/3 power transform
eps4 <- scale(rnorm(n)^2, center  = TRUE)      # squared normal, centered
# Generate variables
x2 <- eps2
x1 <- 0.9*x2 + eps1
x3 <- 0.8*x2 + eps3
x4 <- -x1 - 0.9*x3 + eps4
# Data matrix
X <- cbind(x1, x2, x3, x4)
# True DAG in matrix form
trueDAG <- cbind(
  x1 = c(0,1,0,0),
  x2 = c(0,0,0,0),
  x3 = c(0,1,0,0),
  x4 = c(1,0,1,0)
)
# Causal relations:
## x2 → x1,  x2 → x3,  x1 → x4,  x3 → x4
# Apply LiNGAM (with verbose output)
res1 <- lingam(X, verbose = TRUE)
res2 <- lingam(X, verbose = 2)  # includes fastICA details
# Sanity check: results coincide
stopifnot(identical(res1, res2))
# Show true DAG
cat("【True 】 DAG:\n")
show(trueDAG)
# Show estimated DAG
cat("【Estimated 】 DAG:\n")
show(as(res1, "amat"))
```

Readers are encouraged to run the program themselves (Problem 64).

From the results, the structure estimated by LiNGAM broadly agrees with the true DAG. In particular, x_4 is correctly identified as being influenced by x_1 and x_3, showing that LiNGAM works effectively under the non-Gaussianity assumption. By adjusting the `verbose` option, one can also inspect details of the LiNGAM computations and the fastICA procedure. ■

6.6 Handling Confounding via the Shortest-Path Problem

In LiNGAM, it is assumed in (6.1) and (6.2) that at least one of $e_1 \perp\!\!\!\perp e_2$ or $e'_1 \perp\!\!\!\perp e'_2$ holds. How should we estimate the causal ordering when neither holds (i.e., when confounding is present)?

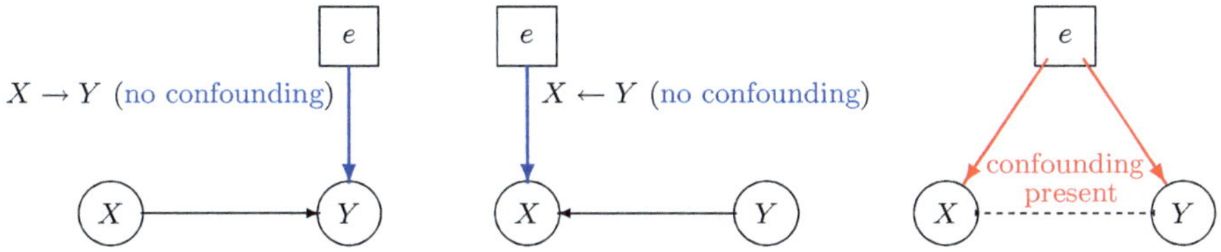

Fig. 6.6 When noise affects only a single variable (shown in blue), LiNGAM can identify the causal direction as either $X \to Y$ or $Y \to X$, provided that at least one of X or Y is non-Gaussian. When noise affects multiple variables (shown in red), the causal direction cannot be identified

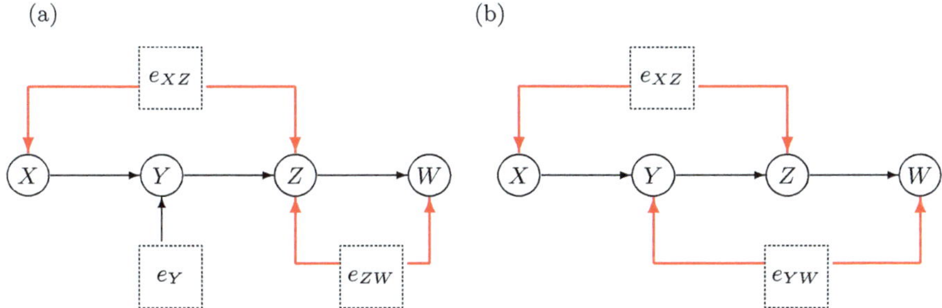

Fig. 6.7 In (**a**), $x^n \perp\!\!\!\perp y^n_x$ and $y^n_x \perp\!\!\!\perp z^n_{xy}$ hold, but $x^n \perp\!\!\!\perp z^n_{xy}$ does not. Therefore, only $X \to Y \to Z$ can be identified. In (**b**), since $x^n \perp\!\!\!\perp y^n_x$, $y^n_x \perp\!\!\!\perp z^n_{xy}$, and $z^n_{xy} \perp\!\!\!\perp w^n_{xyz}$ all hold, $X \to Y \to Z \to W$ can be identified

In this book, *confounding* refers to situations in which noise affects multiple random variables, making the causal direction difficult to identify, and such noise is called a *confounding factor* (Fig. 6.6).

Example 49 In Fig. 6.7a, when we compute x^n, y^n_x, z^n_{xy}, $x^n \perp\!\!\!\perp z^n_{xy}$ does not hold because of the confounding factor e_{XZ}. However, if $x^n \perp\!\!\!\perp y^n_x$ and $y^n_x \perp\!\!\!\perp z^n_{xy}$ hold, then $X \to Y \to Z$ can be identified. On the other hand, because of the confounding factor e_{ZW}, $z^n_{xy} \perp\!\!\!\perp w^n_{xyz}$ does not hold. In Fig. 6.7b, if

$$x^n \perp\!\!\!\perp y^n_x, \qquad y^n_x \perp\!\!\!\perp z^n_{xy}, \qquad z^n_{xy} \perp\!\!\!\perp w^n_{xyz}$$

all hold, then $X \to Y \to Z \to W$ can be identified ■

From Example 49, we see that in LiNGAM, one can identify the ordering among variables by testing only the independence of residuals between adjacent variables. Thus, even if confounding factors exist, as long as the confounding does not occur between adjacent variables, the standard LiNGAM procedure is applicable.

Proposition 22 (Wang–Drton, 2023 [41]) *Even if there is a confounding factor affecting a pair of variables, the standard LiNGAM procedure is applicable as long as the confounding does not occur between adjacent variables.*

However, in practice, confounding often arises between adjacent variables, so Proposition 22 is of limited practical use. For $p = 2$, it is common to compare the plausibility of $X \to Y$ vs. $Y \to X$ using HSIC values or by comparing mutual information.

How, then, should we proceed in the general case of p variables?

To address this, let us return to the independent component analysis (ICA) problem. In fastICA, the goal is to maximize negentropy. In the original paper by Aapo Hyvärinen [9], it is argued that when extracting multiple components, one should maximize the sum

$$\text{constant} - \sum_{i=1}^{p} h(f_i).$$

Using the notation up to this point, for random variables $X_1, \ldots, X_p$, if we denote by $f_{1,\ldots,p}(e_1, \ldots, e_p)$ the joint probability density function of the residuals $e_1, \ldots, e_p$ given the upstream variables, then

$$KL(e_1, \ldots, e_p) := \mathbb{E}\left[\log \frac{f_{1,\ldots,p}(e_1, \ldots, e_p)}{f_1(e_1) \cdots f_p(e_p)}\right] \tag{6.17}$$

$$= \int \cdots \int f_{1,\ldots,p}(e_1, \ldots, e_p) \log \frac{f_{1,\ldots,p}(e_1, \ldots, e_p)}{f_1(e_1) \cdots f_p(e_p)} de_1 \cdots de_p$$

corresponds to minimizing this quantity. Note that this distribution depends on the ordering of $X_1, \ldots, X_p$.

In ICA, (6.17) is referred to as the mutual information of p variables. It can be viewed as the Kullback–Leibler divergence between $f_{1,\ldots,p}$ and the product $f_1 \cdots f_p$ of its marginals $f_1, \ldots, f_p$.

From the foregoing discussion, the absence of confounding is equivalent to the value of (6.17) being zero. LiNGAM can determine a variable ordering by the same procedure regardless of whether confounding is present, and the value of (6.17) varies with the ordering.

Below, we define the magnitude of confounding by the value of (6.17).

When there is no confounding, $P(e_1, \ldots, e_p) = P(e_1) \cdots P(e_p)$ holds and $KL(e_1, \ldots, e_p) = 0$. This is what LiNGAM assumes for a given variable ordering.

Note that mutual information $I(X, Y)$ is nonnegative and satisfies

$$I(X, Y) = 0 \Longleftrightarrow X \perp\!\!\!\perp Y.$$

Therefore, confounding can be expressed as the following sum of mutual information:

$$KL(e_1, \ldots, e_p) = E\left[\log \frac{P(e_1, \ldots, e_p)}{P(e_1)P(e_2, \ldots, e_p)}\right] + E\left[\log \frac{P(e_2, \ldots, e_p)}{P(e_2) \cdots P(e_3, \ldots, e_p)}\right]$$

$$+ \cdots + E\left[\log \frac{P(e_{p-1}, e_p)}{P(e_{p-1})P(e_p)}\right]. \tag{6.18}$$

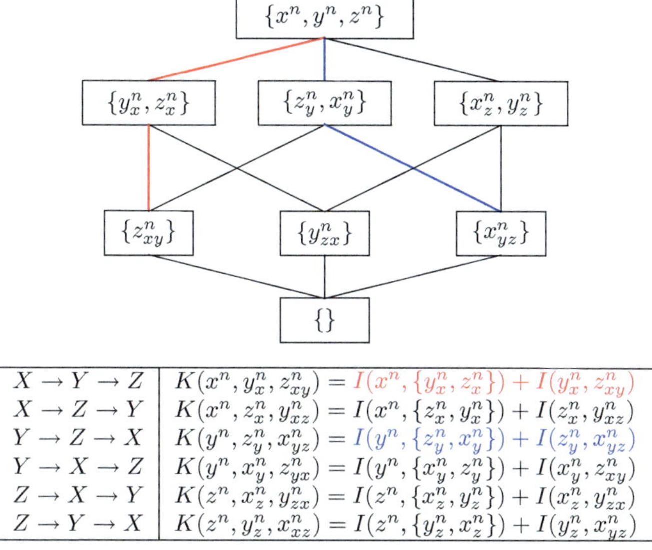

$X \to Y \to Z$	$K(x^n, y^n_x, z^n_{xy}) = I(x^n, \{y^n_x, z^n_x\}) + I(y^n_x, z^n_{xy})$
$X \to Z \to Y$	$K(x^n, z^n_x, y^n_{xz}) = I(x^n, \{z^n_x, y^n_x\}) + I(z^n_x, y^n_{xz})$
$Y \to Z \to X$	$K(y^n, z^n_y, x^n_{yz}) = I(y^n, \{z^n_y, x^n_y\}) + I(z^n_y, x^n_{yz})$
$Y \to X \to Z$	$K(y^n, x^n_y, z^n_{yx}) = I(y^n, \{x^n_y, z^n_y\}) + I(x^n_y, z^n_{xy})$
$Z \to X \to Y$	$K(z^n, x^n_z, y^n_{zx}) = I(z^n, \{x^n_z, y^n_z\}) + I(x^n_z, y^n_{zx})$
$Z \to Y \to X$	$K(z^n, y^n_z, x^n_{xz}) = I(z^n, \{y^n_z, x^n_z\}) + I(y^n_z, x^n_{yz})$

Fig. 6.8 There are six paths corresponding to the six possible orders. Along each path (order), the total distance from the top $\{x^n, y^n, z^n\}$ to the bottom $\{\}$ is compared

Furthermore, since mutual information is nonnegative, the following equivalences hold (Fig. 6.8):

$$\begin{aligned}\text{No confounding} &\iff KL(e_1, \ldots, e_p) = 0\\ &\iff I(e_1, \{e_2, \ldots, e_p\}) = \cdots = I(e_{p-1}, e_p) = 0\\ &\iff e_1, \ldots, e_p \text{ are mutually independent.}\end{aligned}$$

We now consider minimizing the sum (6.18) of mutual information. To that end, we prepare an ordered graph for finding the optimal ordering using the *shortest-path problem.*

Suppose there are three variables X, Y, Z ($p = 3$). To compare the six possible orderings, we construct a graph as in Fig. 6.8. For example, to compare the orderings $X \to Y \to Z$ and $Y \to Z \to X$, we compute the total lengths along the red and blue paths.

Given data $\{x^n, y^n, z^n\}$ as input, we compute the length of each edge (the estimated mutual information), assuming $I_n(\{x^n_{yz}\}, \{\}) = I_n(\{y^n_{zx}\}, \{\}) = I_n(\{z^n_{xy}\}, \{\}) = 0$.

Treat these estimated mutual information as distances. For each node v, compute the path length $d(v)$ from the vertex $\{x^n, y^n, z^n\}$ to v as the sum of edge lengths along a path. If multiple paths to a node exist, select the shortest among them and record it for that node.

For example, the total length of the path $\{x^n, y^n, z^n\} \to \{y^n_x, z^n_x\} \to \{z^n_{xy}\} \to \{\}$ is

$$I_n(x^n, \{y^n_x, z^n_x\}) + I_n(y^n_x, z^n_{xy}) + 0 = I_n(x^n, \{y^n_x, z^n_{xy}\}) + I_n(y^n_x, z^n_{xy}),$$

which corresponds to the estimated *KL information* of the noise set $\{e_1, e_2, e_3\}$ in the model $X = e_1, Y = aX + e_2, Z = bX + cY + e_3$. Our goal is to find the shortest path from $\{x^n, y^n, z^n\}$ to $\{\}$.

As shown in Figure 5.8, first compute the lengths of the edges from the vertex $\{X, Y, Z\}$ to $\{Y, Z\}, \{Z, X\}, \{X, Y\}$:

$$d(\{Y, Z\}) := I_n(x^n, \{y^n_x, z^n_x\}),$$

$$d(\{Z, X\}) := I_n(y^n, \{z^n_y, x^n_y\}),$$

$$d(\{X, Y\}) := I_n(z^n, \{x^n_z, y^n_z\}).$$

At this point, move $\{X, Y, Z\}$ to CLOSE and add $\{Y, Z\}, \{Z, X\}, \{X, Y\}$ to OPEN (Fig. 6.9a).

If $d(\{Y, Z\})$ is the smallest, compute $I_n(y^n_x, z^n_{xy})$ and $I_n(z^n_x, y^n_{zx})$ and update as follows:

$$d(\{Z\}) := d(\{Y, Z\}) + I_n(y^n_x, z^n_{xy}), \tag{6.19}$$

$$d(\{Y\}) := d(\{Y, Z\}) + I_n(z^n_x, y^n_{zx}).$$

Then move $\{Y, Z\}$ to CLOSE and add $\{Z\}, \{Y\}$ to OPEN (Fig. 6.9b).

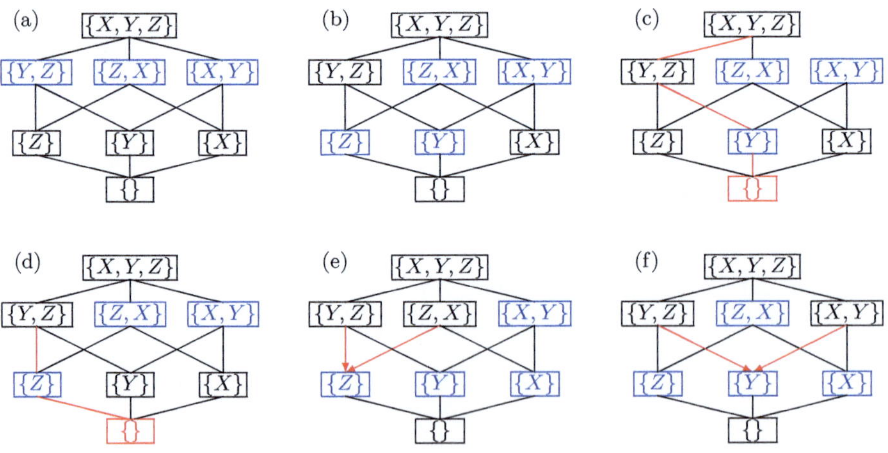

Fig. 6.9 Searching for causal orderings via the shortest-path problem

Next, among $d(\{Y\})$, $d(\{Z\})$, $d(\{Z, X\})$, and $d(\{X, Y\})$, if $d(\{Y\})$ is minimal, the shortest path is determined as in Fig. 6.9c, and the variable ordering is $X \to Z \to Y$. If $d(\{Z\})$ is minimal, the shortest path is as in Fig. 6.9d, and the ordering is $X \to Y \to Z$.

If in Fig. 6.9b $d(\{Z, X\})$ is minimal, compute $I_n(z^n_y, x^n_{yz})$ and $I_n(x^n_y, z^n_{xy})$ and update:

$$d(\{X\}) := d(\{Z, X\}) + I_n(z^n_y, x^n_{yz})$$

$$d(\{Z\}) := d(\{Z, X\}) + I_n(x^n_y, z^n_{xy}) \ . \tag{6.20}$$

Then CLOSE $\{Z, X\}$ and OPEN $\{X\}$ and $\{Z\}$. Since the values in (6.19) and (6.20) may conflict, if the value from (6.20) is smaller, replace (6.19) with (6.20) (Fig. 6.9e).

Finally, if in Fig. 6.9b $d(\{X, Y\})$ is minimal, proceed to Fig. 6.9f. Although the values of $d(\{Y\})$ conflict, the shorter of the paths from $\{X, Y, Z\}$ to $\{Y\}$ is selected.

By repeating this procedure, we ultimately obtain the distance $d(\{\})$ and find the shortest path from $\{X, Y, Z\}$ to $\{\}$. The details of this procedure (LiNGAM-MMI) are given below. It takes $DATA$ as input and outputs $SHORTEST\_PATH$.

This method applies not only to $p = 3$ but to any p.

[Determining causal order via shortest-path search]

1. Initialization:
   - $OPEN := \{TOP\}$, $CLOSE := \{\}$, $path(TOP) := ()$, $r(TOP) := DATA$, $d(TOP) = 0$
2. Loop:
   (a) Among nodes v in $OPEN$, select the one with the smallest $d(v)$ and move it to $CLOSE$.
   (b) For the nodes $v_1, \ldots, v_m$ adjacent to v, for each v_i:
      i. If $v_i \notin OPEN$, compute the residual $r(v_i)$ from $r(v)$.
      ii. Using $r(v)$ and $r(v_i)$, compute the mutual information mi.
      iii. If $v_i \notin OPEN$ or ($v_i \in OPEN$ and $d(v) + mi < d(v_i)$), then:
         - Set $d(v_i) = d(v) + mi$ and $path(v_i) = append(path(v), v_i)$.
      iv. If $v_i \notin OPEN$, add v_i to $OPEN$.
   (c) If BOTTOM $\in$ $OPEN$, output $SHORTEST\_PATH = append(path(v), \{\})$ and terminate.

Here, TOP and $BOTTOM$ denote the top and bottom nodes of the graph, respectively, and for nodes $u_1, u_2, \cdots, u_{s+1}$, we define

$$append((u_1, \cdots, u_s), u_{s+1}) := (u_1, \cdots, u_s, u_{s+1}).$$

The following proposition holds for this method. The approach of Wang–Drton [41], designed to handle a specific type of confounding, requires exponential time in p even when there is no confounding. In practice, from data alone we cannot know a priori whether confounding is present. Proposition 23 asserts that although computation time may be substantial when confounding is present, at least when there is no confounding, the computation finishes in polynomial time in p.

Proposition 23 (Suzuki–Yang 2024 [39]) *The method based on the shortest-path problem identifies the causal ordering with probability 1 after*

$$p + (p-1) + \cdots + 1 = \frac{p(p+1)}{2} \tag{6.21}$$

mutual information computations, provided that the confounding factors do not act between adjacent variables, and the LiNGAM solution is unique.

Proof At TOP, compute p mutual information; the minimum is 0. For the newly opened $p-1$ vertices in $OPEN$, compute the mutual information; again there exists one whose value is 0. Repeat this operation until reaching $BOTTOM$. The total number of mutual information computations, as well as the number of comparisons, is $p(p+1)/2$. ■

Minimizing (6.18) is equivalent to selecting, among the $p!$ models without confounding, the one whose Kullback–Leibler information is minimal.

There is also an advantage even when no confounding exists. Typically, only data are given, and it is more appropriate to view the task as estimation rather than exact identification of the causal ordering. In such cases, when one greedily determines the order from upstream using HSIC or similar criteria, one may pick an incorrect variable. Applying the shortest-path formulation amounts to performing a global search for the optimal ordering.

For numerical experiments and other details, see [39].

Appendix

Proof of (6.9)

First, we have

$$\left(y^n - \frac{c(x^n, y^n)}{v(x^n)}x^n\right)_z = \frac{c\left(y^n - \dfrac{c(x^n, y^n)}{v(x^n)}x^n, z^n\right)}{v\left(y^n - \dfrac{c(x^n, y^n)}{v(x^n)}x^n\right)}\left\{y^n - \frac{c(x^n, y^n)}{v(x^n)}x^n\right\}$$

$$c\left(y^n - \frac{c(x^n, y^n)}{v(x^n)}x^n, z^n\right) = c(y^n, z^n) - \frac{c(x^n, y^n)c(x^n, z^n)}{v(x^n)}$$

$$= \frac{v(x^n)c(y^n, z^n) - c(x^n, y^n)c(x^n, z^n)}{v(x^n)}$$

$$v\left(y^n - \frac{c(x^n, y^n)}{v(x^n)}x^n\right) = v(y^n) - 2\frac{c(x^n, y^n)^2}{v(x^n)} + \frac{c(x^n, y^n)^2}{v(x^n)}$$

$$= \frac{v(x^n)v(y^n) - c(x^n, y^n)^2}{v(x^n)}$$

Therefore, z^n_{xy} is given as follows, yielding (6.9)

$$\begin{aligned} z^n_{xy} =& z^n_x - (y^n_x)_z = (z^n - \frac{c(x^n, z^n)}{v(x^n)}) - (y^n - \frac{c(x^n, y^n)}{v(x^n)})_z \\ =& z^n - \frac{c(x^n, z^n)}{v(x^n)}x^n - \frac{v(x^n)c(y^n, z^n) - c(x^n, y^n)c(x^n, z^n)}{v(x^n)v(y^n) - c(x^n, y^n)^2}\{y^n - \frac{c(x^n, y^n)}{v(x^n)}x^n\} \\ =& z^n - \frac{v(y^n)c(x^n, z^n) - c(x^n, y^n)c(y^n, z^n)}{v(x^n)v(y^n) - c(x^n, y^n)^2}x^n \\ & - \frac{v(x^n)c(y^n, z^n) - c(x^n, y^n)c(x^n, z^n)}{v(x^n)v(y^n) - c(x^n, y^n)^2}y^n \end{aligned} \tag{6.22}$$

■

Proof of Proposition 19

Since the random variables

$$Y_1 = \sum_{i=1}^{m} \alpha_i X_i, \qquad Y_2 = \sum_{i=1}^{m} \beta_i X_i,$$

are independent, their two-dimensional characteristic function can be written, by (2.26), as $\Phi_{Y_1Y_2}(u, v) = \Phi_{Y_1}(u)\Phi_{Y_2}(v)$. Using the independence of $X_1, \ldots, X_N$, we obtain

$$\begin{aligned} \Phi_{Y_1Y_2}(u, v) &= \mathbb{E}_{Y_1Y_2}\left[\exp\left\{i\left(u\sum_{j=1}^{m}\alpha_j X_j + v\sum_{j=1}^{m}\beta_j X_j\right)\right\}\right] \\ &= \mathbb{E}_{X_1\cdots X_m}\left[\exp\left\{i\sum_{j=1}^{m}\left(\alpha_j u + \beta_j v\right)X_j\right\}\right] \\ &= \prod_{j=1}^{m}\mathbb{E}_{X_j}\left[\exp\left\{i\left(\alpha_j u + \beta_j v\right)X_j\right\}\right] = \prod_{j=1}^{m}\Phi_{X_j}(\alpha_j u + \beta_j v) \end{aligned}$$

Hence, identically,

$$\psi_{X_1}(\alpha_1 u + \beta_1 v) + \cdots + \psi_{X_m}(\alpha_m u + \beta_m v) = \psi_{Y_1}(u) + \psi_{Y_2}(v) \tag{6.23}$$

holds, where we set $\phi_{X_i}(\cdot) := \log \Phi_{X_i}(\cdot)$, $\phi_{Y_j}(\cdot) := \log \Phi_{Y_j}(\cdot)$.[2]

Next, move those terms on the left-hand side to the right-hand side in which at least one of α_j, β_j equals 0. Denote the resulting form of (6.23) by

$$\sum_{j=1}^{n} \Delta_j^{(0)}(u, v) = A_0(u) + B_0(v) \tag{6.24}$$

(with $n \leq m$). If initially $\alpha_j \beta_j \neq 0$ holds for all j, then

$$\Delta_j^{(0)}(u, v) = \psi_{X_j}(\alpha_j u + \beta_j v), \qquad A_0(u) = \phi_{Y_1}(u), \qquad B_0(u) = \phi_{Y_2}(u)$$

(with $m = n$). Now vary (u, v) to $(u+r, v+s)$ so that the value of $\alpha_1 u+\beta_1 v$ remains unchanged, i.e., for (r, s) satisfying $\alpha_1 r + \beta_1 s = 0$, and subtract the original (6.24) from the transformed one. Defining

$$\begin{aligned} \Delta_j^{(1)}(u, v) &:= \Delta_j^{(0)}(u + r, v + s) - \Delta_j^{(0)}(u, v) \\ A_1(u) &:= A_0(u + r) - A_0(u) \\ B_1(u) &:= B_0(v + s) - B_0(v) \end{aligned}$$

we obtain

$$\sum_{j=2}^{n} \Delta_j^{(1)}(u, v) = A_1(u) + B_1(v) .$$

Next, by changing u, v so that the value of $\alpha_2 u + \beta_2 v$ is unchanged, and omitting technical details, we get

$$\sum_{j=3}^{n} \Delta_j^{(2)}(u, v) = A_2(u) + B_2(v) .$$

Repeating this procedure yields

$$\Delta_n^{(n-1)}(u, v) = A_{n-1}(u) + B_{n-1}(v) .$$

[2] A characteristic function $\Psi(\cdot)$ satisfies $\Psi(0) = 1$ and is continuous; hence, for t in a neighborhood of the origin, $\Psi(t) \neq 0$ (so $\log \varphi(t) < \infty$).

Furthermore, changing only u to $u+r$ and subtracting both sides, and then changing only v to $v+s$ and subtracting both sides, we finally get

$$\Delta_n^{(n+1)}(u, v) = 0\,.$$

That is, since $\Delta_j^{(0)}$ becomes identically zero after $n+1$ differentiations, it is a polynomial of degree at most n. Hence, for each j with $\alpha_j\beta_j \neq 0$, ψ_{X_j} is a polynomial of degree at most n.

Here, we use the following lemma (see the appendix for a proof). See Section 3.3 "Marcinkiewicz's Theorem" of [16]. A clear proof is also given in Section 2.5 "Cramer and Marcinkiewicz theorems" of [12].

Lemma 2 (Marcinkiewicz, 1938 [17]) *If a characteristic function can be written as $\Psi(t) = e^{P(t)}$ for a polynomial $P(t)$, then the degree of $P(t)$ is at most 2.*

If the degree is 2 and the leading coefficient is positive, then the logarithm of the characteristic function is unbounded. Also, since the characteristic function equals 1 at 0 and its logarithm is 0 there, the constant term of the polynomial must be 0. Therefore, for any k with $\alpha_k\beta_k \neq 0$, we can write $\phi_{X_k}(t) = i\mu t - \sigma^2 t^2/2$. If $\sigma^2 = 0$, then X_k is constant, so in fact $\sigma^2 > 0$. Hence, $\Psi_k(\cdot)$ is the characteristic function of a normal distribution, and X_k is normally distributed. ■

Proof of Proposition 21

For example, suppose $X_1 \to X_2 \to X_3 \to X_4 \to X_5 \to X_6$ is the correct order, and their joint density factorizes as

$$f(x_1, x_2, x_3, x_4, x_5, x_6) = f(x_1, x_2)\frac{f(x_3, x_4, x_5)f(x_3, x_4, x_6)}{f(x_3, x_4)}\,.$$

Then orders such as

$$X_3 \to X_4 \to X_5 \to X_6 \to X_1 \to X_2, \qquad X_1 \to X_2 \to X_3 \to X_4 \to X_6 \to X_5$$

are also valid. Below, we consider independent groups of variables such as $\{X_1, X_2\}$ and $\{X_3, X_4, X_5, X_6\}$ independently of each other, and within them parallel groups such as $\{X_3, X_4, X_5\}$ and $\{X_3, X_4, X_6\}$, and we deem the overall system identifiable if all of these are identifiable. Thus, without loss of generality, we assume that in the SEM

$$X_i = \sum_{j=1}^{i-1} \beta_{i,j} X_j + e_i$$

the coefficients $\beta_{i,j}$ are nonzero ($j = 1, \ldots, i-1, i = 1, \ldots, p$).

For each $i = 2, \ldots, p$, since $e_1 \perp\!\!\!\perp X_j$ $(j = 2, \ldots, i-1)$ and $e_1 \perp\!\!\!\perp e_i$, it follows that $X_1 = e_1$ and the residual

$$X_i - \beta_{i,1}X_1 = \sum_{j=2}^{i-1} \beta_{i,j}X_j + e_i$$

are independent.

Conversely, suppose there exists a constant $\alpha \neq 0$ such that $X_i \perp\!\!\!\perp (e_1 - \alpha X_i)$. That is, assume there is a variable X_i upstream of $X_1 = e_1$ such that X_i is independent of the residual $X_1 - \alpha X_i$ for some $\alpha \neq 0$ (if this held with $\alpha = 0$, then $X_i \perp\!\!\!\perp X_1$, contradicting the factorization assumption). Now, taking $i \geq 3$ (excluding $i = 2$ since it has already been proved), note that there exist constants $\gamma_1, \ldots, \gamma_i$ such that

$$X_i = \sum_{j=1}^{i-2} \gamma_j e_j + \beta_{i,i-1}e_{i-1} + e_i$$

$$e_1 - \alpha X_i = (1 - \alpha\gamma_1)e_1 - \alpha \sum_{j=2}^{i-2} \gamma_j e_j - \alpha\beta_{i,i-1}e_{i-1} - \alpha e_i$$

For instance, when $i = 3$,

$$X_3 = (\beta_{3,1} + \beta_{3,2}\beta_{2,1})e_1 + \beta_{3,2}e_2 + e_3,$$

so $\gamma_1 = \beta_{3,1} + \beta_{3,2}\beta_{2,1}$ and $\gamma_2 = \beta_{3,2}$. Since $\alpha, \beta_{i,i-1} \neq 0$, applying Proposition 19 implies that e_{i-1} and e_i must both be Gaussian, which contradicts the assumption. ∎

Problems 54–69

54. Explain why the fact that swapping x^n and y^n does not change the value of (6.9) implies that $z^n_{xy} = z^n_{yx}$.
55. Fill in the blanks below to construct the function `search.p.val`, which explores variable ordering based on p-values.

```
search.p.val <- function(index, Z) {
  # If the length of index is less than 2, return index as is
  if (length(index) < 2) return(index)
  # Initialize: set the maximum p-value
  max <- -1
  # Loop over each element of index
  for (i in index) {
    # Move i to the beginning, followed by the rest in order
```

```
      index.2 <- c(i, setdiff(index, i))
      # Run LiNGAM.2 function to obtain p-value
      result <- ## Blank (1) ##
      # Update the maximum p-value and record Z and i
      if ( ## Blank (2) ##) {
        max <- ## Blank (3) ##
        W <- Z
        W[index.2] <- result$Z  # Save updated Z after applying p-value
        k <- i
      }
    }
    # Recursively call with updated W and combine the results
    index.3 <- ## Blank (4) ##
    return(c(k, index.3))
}
```

56. As in Example 45, apply `search.HSIC` to the dataset `crime.txt` included in the CRAN `glmnet` package, and determine the causal ordering of the seven variables. The meaning of each variable is as follows:

X1	Overall crime rate per 1 million population
X2	Violent crime rate per 100,000 population
X3	Annual police budget per capita
X4	Percentage of people aged 25 and older who completed 4 years of high school
X5	Percentage of people aged 16–19 not attending or not having completed high school
X6	Percentage of people aged 18–24 enrolled in college
X7	Percentage of people aged 25 and older who graduated from a 4-year college

57. In the discussion right after Proposition 19, prove that

$$a = 0 \Longleftrightarrow a' = 0 \Longleftrightarrow X \perp\!\!\!\perp Y.$$

Also explain why, as long as $\sigma_2^2 > 0$, it is impossible for $aa' = 1$ to hold.

58. In (6.10), under the assumption that $cov(e_1, e_2) = cov(e'_1, e'_2) = 0$, show that if $e_1 \perp\!\!\!\perp e_2$ and e_1, e_2 follow the Gaussian distribution, then $e'_1 \perp\!\!\!\perp e'_2$ and e'_1, e'_2 also follow the Gaussian distribution.

59. Consider identifiability when the true order $X \to Y \to Z$ (residuals: e_1, e_2, e_3) is misinterpreted as $Y \to Z \to X$ (residuals: e'_1, e'_2, e'_3).

(a) Show that when

$$\begin{cases} X = e_1 \\ Y = aX + e_2 \\ Z = bX + cY + e_3 \end{cases}, \qquad \begin{cases} Y = e'_1 \\ Z = a'Y + e'_2 \\ X = b'Y + c'Z + e'_3 \end{cases}$$

the residuals satisfy

$$\begin{aligned} e_1' &= ae_1 + e_2 \\ e_2' &= \{b + a(c - a')\}e_1 + (c - a')e_2 + e_3 \\ e_3' &= \{1 - bc' - a(b' + cc')\}e_1 - (b' + cc')e_2 - c'e_3 \end{aligned}$$

(b) Assuming no correlation among e_1', e_2', e_3', show that a', b', c' can be written in terms of a, b, c, and derive

$$\begin{aligned} e_1' &= ae_1 + e_2 \\ e_2' &= \frac{b\sigma_2^2}{a^2\sigma_1^2 + \sigma_2^2}e_1 - \frac{ab\sigma_1^2}{a^2\sigma_1^2 + \sigma_2^2}e_2 + e_3 \\ e_3' &= \frac{\sigma_2^2\sigma_1^2}{b^2\sigma_1^2\sigma_2^2 + (a^2\sigma_1^2 + \sigma_2^2)\sigma_3^2}e_1 - \frac{a\sigma_1^2\sigma_3^2}{b^2\sigma_1^2\sigma_2^2 + (a^2\sigma_1^2 + \sigma_2^2)\sigma_3^2}e_2 \\ &\quad - \frac{b\sigma_1^2\sigma_2^2}{b^2\sigma_1^2\sigma_2^2 + (a^2\sigma_1^2 + \sigma_2^2)\sigma_3^2}e_3 \end{aligned}$$

(c) Show that when $a, b, c \neq 0$, if at least one of e_1, e_2, e_3 is non-Gaussian, then no pair among e_1', e_2', e_3' is independent.

(d) Show that if $Y = ae_1 + e_2$ is non-Gaussian, then at least one of e_1, e_2 is non-Gaussian; if $Z = be_1 + c(ae_1 + e_2) + e_3$ is non-Gaussian, then at least one of e_1, e_2, e_3 is non-Gaussian.

(e) Show that if at least one of X, Y, Z is non-Gaussian, then e_1', e_2', e_3' cannot be mutually independent. In other words, misidentifying $X \to Y \to Z$ as $Y \to Z \to X$ does not occur.

[Hint] True model: $X = e_1,\ Y = aX + e_2,\ Z = bX + cY + e_3$. Mis-specified model: $Y = e_1',\ Z = a'Y + e_2',\ X = b'Y + c'Z + e_3'$. Since e_1' is Y, it equals $ae_1 + e_2$; e_2' is $Z - a'Y$; e_3' is $X - b'Y - c'Z$. Choose a', b', c' so that covariances vanish. Focus on independence of residuals: identifiability fails if Gaussianity is assumed, but not otherwise.

60. Consider misidentification of the true order $X \to Y \to Z$ (residuals: e_1, e_2, e_3) as either $Y \to X \to Z$ (residuals: e_1', e_2', e_3') or $X \to Z \to Y$ (residuals: e_1'', e_2'', e_3'').

(a) Show that they can be written as

$$\begin{cases} Y = e_1' \\ X = a'Y + e_2' \\ Z = b'Y + c'X + e_3' \end{cases}, \quad \begin{cases} X = e_1'' \\ Z = a''X + e_2'' \\ Y = b''X + c''Z + e_3'' \end{cases}$$

(b) By requiring pairwise covariances of $\{e_1', e_2', e_3'\}$ to vanish and eliminating a', b', c' and similarly eliminating a'', b'', c'' for $\{e_1'', e_2'', e_3''\}$, show that the following holds:

$$\begin{cases} e_1' = ae_1 + e_2 \\ e_2' = \dfrac{\sigma_2^2}{a^2\sigma_1^2 + \sigma_2^2} e_1 - \dfrac{a\sigma_1^2}{a^2\sigma_1^2 + \sigma_2^2} e_2 \\ e_3' = e_3 \end{cases}, \quad \begin{cases} e_1'' = e_1 \\ e_2'' = ce_2 + e_3 \\ e_3'' = \dfrac{\sigma_3^2}{c^2\sigma_2^2 + \sigma_3^2} e_2 \\ \quad - \frac{c\sigma_2^2}{c^2\sigma_2^2+\sigma_3^2} e_3 \end{cases}$$

(c) Show that when e_3 is non-Gaussian and e_1, e_2 are Gaussian, misidentification as $Y \to X \to Z$ is possible.
(d) Show that when e_1 is non-Gaussian and e_2, e_3 are Gaussian, misidentification as $X \to Z \to Y$ is possible.

[Hint] True order: $X = e_1,\ Y = aX + e_2,\ Z = cY + e_3$. Mis-order 1: $Y = e_1',\ X = a'Y + e_2',\ Z = b'Y + c'X + e_3'$. Mis-order 2: $X = e_1'',\ Z = a''X + e_2'',\ Y = b''X + c''Z + e_3''$. Choose coefficients so that covariances vanish and express e_1', e_2', e_3' in terms of e_1, e_2, e_3. If e_3 is non-Gaussian, focus on residuals such as e_3' that involve it. By Darmois–Skitovich, if at least one is non-Gaussian, linear independence is broken. Thus, residuals of mis-specified models are not independent, preventing misidentification.

61. Various measures are used to evaluate deviations from Gaussianity. In addition to mean and variance, skewness and kurtosis are defined as

$$\mathbb{E}_Y[(Y - \mu)^3/\sigma^3],$$

$$\mathbb{E}_Y[(Y - \mu)^4/\sigma^4] - 3,$$

respectively (Fig. 6.10).

(a) Show that these equal 0 under the Gaussian distribution.
(b) Using the notation in this chapter, express skewness and kurtosis in terms of $G(s)$.

[Hint] Skewness measures asymmetry, and kurtosis measures peakedness. A Gaussian distribution $\mathcal{N}(\mu, \sigma^2)$ is symmetric with mean = median = mode, so skewness is 0. The fourth moment (kurtosis) is 0 for the standard normal. Also, in this chapter, ICA uses $G(s)$ as a measure of non-Gaussianity. Skewness and kurtosis can be written in the form $\mathbb{E}[G(s)]$.

62. (a) Show that the differential entropy of $X \sim N(0, \sigma^2)$ equals the right-hand side of (6.16).
(b) Show that assumption (6.15) implies inequality (6.16).

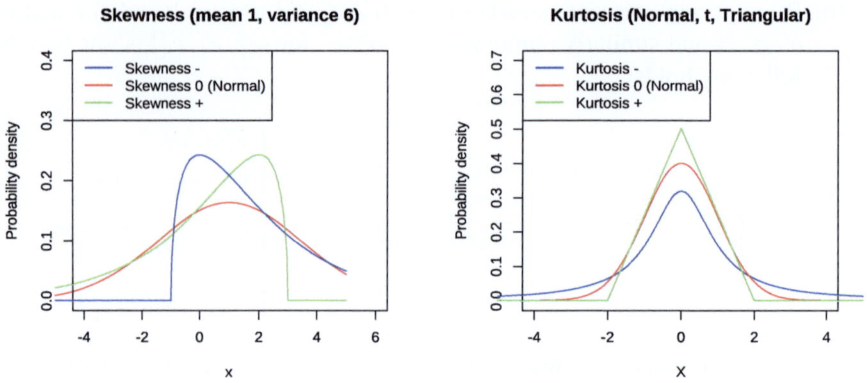

Fig. 6.10 Skewness (left) and kurtosis (right). For the Gaussian distribution (both red curves), skewness and kurtosis are zero

63. Construct an R program to reproduce Fig. 6.5 by filling in the blanks.

```
library(fastICA)
S <- cbind(
  sin((1:1000) / 20),  # First signal: sine wave
  rep((((1:200) - 100) / 100), 5)  # Second signal: repeated linear trend
)
A <- matrix(c(0.291, 0.6557, -0.5439, 0.5572), 2, 2)
X <- S %*% A
a <- fastICA(X, 2, alg.typ = "parallel", fun = "logcosh", alpha = 1, method = "R", row.norm = FALSE, maxit = 200, tol = 0.0001, verbose = TRUE)
# Plot original signals
par(mfcol = c(2, 3))
plot(1:1000, #Blank (1)#, type = "l", main = "Original Signal 1", xlab = "Time", ylab = "Amplitude", col = "blue")
plot(1:1000, #Blank (2)#, type = "l", main = "Original Signal 2", xlab = "Time", ylab = "Amplitude", col = "blue")
# Plot mixed signals
plot(1:1000, #Blank (3)#, type = "l", main = "Mixed Signal 1", xlab = "Time", ylab = "Amplitude", col = "red")
plot(1:1000, X[,2], type = "l", main = "#Blank (4)#", xlab = "Time", ylab = "Amplitude", col = "red")
# Plot estimated independent components by ICA
plot(1:1000, #Blank (5)#, type = "l", main = "Estimated Independent Signal 1", xlab = "Time", ylab = "Amplitude", col = "green")
plot(1:1000, #Blank (6)#, type = "l", main = "Estimated Independent Signal 2", xlab = "Time", ylab = "Amplitude", col = "green")
```

64. Execute the R programs in Examples 47 and 48, and check the outputs.
65. For a permutation matrix P of size p:

(a) List all permutation matrices when $p = 3$, and for each, compute PA, where $A = (a_{i,j})$ is a $p \times p$ matrix.
(b) Explain how to permute the columns of matrix A by using an appropriate matrix.
(c) Explain why there are $p!$ permutation matrices of size p.

[Hint] A permutation matrix P is a $p \times p$ matrix with exactly one 1 in each row and column. Multiplying on the left permutes rows; multiplying on the right by its transpose permutes columns. The number of permutations is $p!$.

66. In the shortest path algorithm, once a vertex is moved to CLOSE, no other path can reach it later. Why? [Hint] Adding a vertex to CLOSE means that it has been finalized as the vertex with the shortest distance among those in OPEN. Any later path reaching it can only be longer or equal, so no update is possible.
67. Prove that the third equality in Eq. (6.22) holds.
68. In the proof of Proposition 20, explain why one may assume $\beta_{i,j} \neq 0$ under the given definition of identifiability, and how Proposition 19 is applied.
69. Generate random three-variable data, then perform independent component extraction using `fastICA` followed by LiNGAM, and also apply direct LiNGAM for order estimation. Compare the results. Through numerical experiments, discuss under what conditions (e.g., noise distribution, sample size) the two estimation methods coincide or diverge.

Chapter 7
Information Criteria and Marginal Likelihood

In this chapter, we discuss information criteria (Akaike Information Criterion—AIC, Bayesian Information Criterion—BIC) and the Bayesian marginal likelihood, which play central roles in structure learning for graphical models.

We first clarify the definitions and objectives of Akaike Information Criterion (AIC) and Bayesian Information Criterion (BIC) and position them as methods for model selection based on the trade-off between goodness of fit and complexity. Using linear regression models and categorical data as examples, we show concrete ways to compute these criteria.

In the second half, we introduce the definition of marginal likelihood within the Bayesian framework and provide examples of likelihood calculations using Jeffreys prior and analytical computations using the inverse-Wishart prior for the normal distribution.

We also use Stirling's formula and Laplace's method to theoretically explain that BIC can be viewed as an approximation of the negative log marginal likelihood. We clarify the relationship between AIC and predictive performance.

These scores are not merely evaluation metrics for models; they can also serve as guidelines for searching for the optimal graphical model among candidate structures. In the next chapter, we will describe in detail score-based structure learning methods that actively leverage such scores. This chapter serves as preparation for that.

7.1 Overview of AIC and BIC

In Chap. 7, to estimate the structure of graphical models, we discover dependencies among random variables from observed data. For that purpose, we use Akaike Information Criterion *(AIC)* and Bayesian Information Criterion (BIC), which are *information criteria.*

J. Suzuki, *Graphical Models and Causal Discovery with R*,
https://doi.org/10.1007/978-981-95-4267-3_7

The quality of a statistical model must account not only for how well the data fit the model but also for how parsimonious the model is. AIC and BIC are defined by

$$AIC = \sum_{i=1}^{n} \{-\log p(x_i \mid \hat{\theta}(x_1, \ldots, x_n))\} + d \tag{7.1}$$

$$BIC = \sum_{i=1}^{n} \{-\log p(x_i \mid \hat{\theta}(x_1, \ldots, x_n))\} + \frac{d}{2} \log n \tag{7.2}$$

Here, $\hat{\theta}$ is the maximum likelihood estimate (see Sect. 2.4) of the parameter θ. Also, d is the *number of parameters*. For example, in linear regression, it refers to the number of regression coefficients (i.e., the number of explanatory variables). AIC and BIC differ only in the second term.

$$\sum_{i=1}^{n} \{-\log p(x_i \mid \hat{\theta}(x_1, \ldots, x_n))\}, \qquad d$$

respectively represent goodness of fit and model parsimony. AIC and BIC weight the first and second terms as 1 vs. 1 and 2 vs. $\log n$, respectively, and select the model with the smaller sum.

It follows immediately from the definitions that the balance between the fit and parsimony scores differs between AIC and BIC. Just as at universities that weigh mathematics more heavily, applicant A may pass and B may fail, while at universities that weigh English more heavily, B may pass and A may fail, AIC places more emphasis on goodness of fit than BIC does and thus tends to select more complex models.

Example 50 Continuing Example 14. Since $\theta_1, \ldots, \theta_\alpha$ sum to 1, they in effect vary in $d = \alpha - 1$ dimensions (Fig. 7.1). Hence, the values of AIC and BIC are

$$AIC = \sum_{j=1}^{\alpha} -k_j \log \frac{k_j}{n} + \alpha - 1 \tag{7.3}$$

$$BIC = \sum_{j=1}^{\alpha} -k_j \log \frac{k_j}{n} + \frac{\alpha - 1}{2} \log n \tag{7.4}$$

■

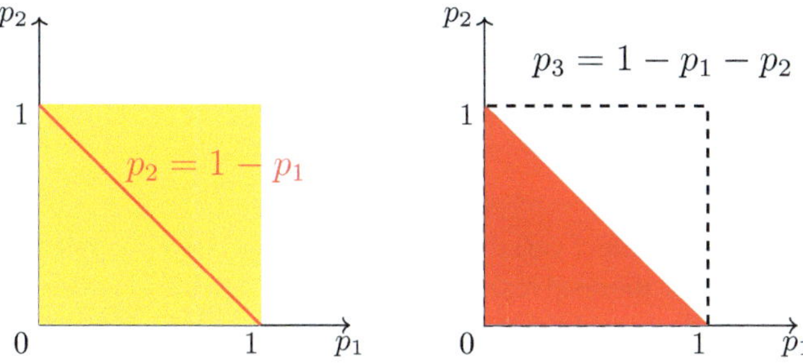

Fig. 7.1 For $\alpha = 2$, even though it appears two-dimensional, (p_1, p_2) can only move along the line segment $p_2 = 1 - p_1$ (left). For $\alpha = 3$, since $p_3 = 1 - p_1 - p_2$ (extending in the direction out of the page) is automatically determined from the values of (p_1, p_2), the interior of the red triangle can be regarded as the parameter space

In R, we can write:

```
AIC.1 <- function(k){
  m <- length(k)
  n <- sum(k)
  return(sum(k*log (n/k))+m-1)
}
```

```
BIC.1 <- function(k){
  m <- length(k)
  n <- sum(k)
  return(sum(k*log (n/k))+(m-1)/2*log(n))
}
```

Example run:

```
> n <- 200
> p <- 0.25
> x <- rbinom(n, 1, p)
> k <- as.vector(table(x))
> AIC.1(k)
[1] 114.5524
> BIC.1(k)
[1] 116.2015
```

Example 51 Continuing Example 15. There are three parameters, $\beta_0, \beta_1, \sigma^2$, so

$$BIC = \frac{n}{2}\log\hat{\sigma}^2 + \frac{n}{2}\log(2\pi e) + \frac{3}{2}\log n$$

is obtained. In general, when there are p slopes (including an intercept), the number of parameters is $p + 1$ (in Example 15, $p = 2$). In that case, (2.33) still holds.

Therefore, BIC is

$$BIC = \frac{n}{2}\log\hat{\sigma}^2 + \frac{n}{2}\log(2\pi e) + \frac{p+1}{2}\log n$$

■

BIC is often used to select variables based on the relative magnitude of its values rather than the values themselves. Thus, one may ignore constants and compare values scaled by a constant. In that sense, in linear regression, it is common to call

$$n\log\hat{\sigma}^2 + p\log n \tag{7.5}$$

"BIC." As long as it is consistent within the discussion whether the intercept is included in the parameter count p, no problem arises.

To compute the concrete value of $\hat{\sigma}^2$ above, we need to obtain $\hat{\beta} \in \mathbb{R}^p$. In matrix form, for observed data $X \in \mathbb{R}^{n\times p}$, $y \in \mathbb{R}^n$, the β that minimizes

$$\|y - X\beta\|^2$$

is given by

$$\hat{\beta} = (X^\top X)^{-1}X^\top y \tag{7.6}$$

Indeed, differentiating with respect to β and setting to zero yields $-X^\top(y-X\beta) = 0$, i.e., $X^\top X\beta = X^\top y$, and as long as $(X^\top X)^{-1}$ exists, (7.6) is the solution. Here $\|\cdot\|^2$ denotes the sum of squares of the n components. Generalizing (2.32), we can compute $\hat{\sigma}^2$ as

$$\hat{\sigma}^2 = \|y - X\hat{\beta}\|^2/n$$

We implement the above in R as follows:

```
sigma2 <- function(X, y){
  n <- length(y)
  beta <- solve(t(X)%*%X) %*% t(X) %*% y
  return( sum((y-X%*%beta)^2) / n)
}
AIC.2 <- function(X, y){
  n <- length(y)
  p <- ncol(X)
  sigma2 <- sigma2(X, y)
  # Substitute into the AIC formula
  value <- (n/2)*log(sigma2) + n/2*log(2*pi*exp(1)) + p + 1
  return(value)
}
BIC.2 <- function(X, y){
  n <- length(y)
  p <- ncol(X)
```

```
  # Compute beta via matrix operations
  sigma2 <- sigma2(X,y)
  # Substitute into the BIC formula
  value <- (n/2)*log(sigma2) + n/2*log(2*pi*exp(1)) + (p+1)/2*log(n)
  return(value)
}
```

In practice, the following AIC and BIC are often used:

```
AIC.3 <- function(X, y) length(y) * log(sigma2(X, y)) + 2*ncol(X)
BIC.3 <- function(X, y) length(y) * log(sigma2(X, y)) + ncol(X)*log(length(y))
```

Example run:

```
> n <- 100
> p <- 3
> X <- matrix(rnorm(n*p), n, p)
> y <- rnorm(n)
> mu <- rnorm(p)
> Sigma <- diag(p)
> AIC.2(X, y)
[1] 137.6937
> BIC.2(X, y)
[1] 142.904
> AIC.3(X, y)
[1] -10.40029
> BIC.3(X, y)
[1] -2.584779
```

The simplified AIC and BIC (`AIC.3`, `BIC.3`) can take negative values.

Finally, in this section, let us empirically verify that AIC tends to select more complex models than BIC.

Example 52 In the Boston dataset (Table 5.1), consider the problem of finding explanatory variables for MEDV (the 14th item) as the response. We ran the following program. We varied the number k of variables from 1 to p and plotted the minimum AIC/BIC for each k. Note that we limited the explanatory variables to those that are continuous. We found that AIC achieves its minimum at $k = 7$, whereas BIC does so at $k = 6$ (Fig. 7.2).

```
library(MASS)
df <- Boston
# Among the 14 variables, keep only discrete explanatory variables and drop the response
X <-as.matrix(df[,c(1,3,5,6,7,8,10,11,12,13)])
y <- df[[14]]
n <- nrow(X)
p <- ncol(X)
X <- cbind(rep(1,n),X) # include intercept
AIC.seq <- NULL
```

```
BIC.seq <- NULL;
for(k in 1:p){
  # combn(S,k) creates a matrix whose columns are the size-k subsets of S
  T <- combn(1:p,k)
  m <- ncol(T)
  # For each k, select the variable combination that minimizes sigma2,
     then compute AIC/BIC for it.
  # Initialize S.min with infinity.
  S.min <- Inf
  for(j in 1:m){
    q <- c(1, T[,j]+1) # intercept (1st col) + selected variables (shifted
      by 1)
    S <- sigma2(X[,q],y)
    if(S < S.min) {
      S.min <- S
      q.min <- q
    }
  }
  AIC.seq <- c(AIC.seq, AIC.3(X[,q.min],y))
  BIC.seq <- c(BIC.seq, BIC.3(X[,q.min],y))
}
plot(1:p, ylim=c(min(AIC.seq),max(BIC.seq)), type="n",
     xlab="Number of variables", ylab="Values of AIC/BIC")
lines(AIC.seq,col="red"); lines(BIC.seq,col="blue")
legend("topright",legend=c("AIC","BIC"), col=c("red","blue"), lwd=1, cex
    =.8)
```

■

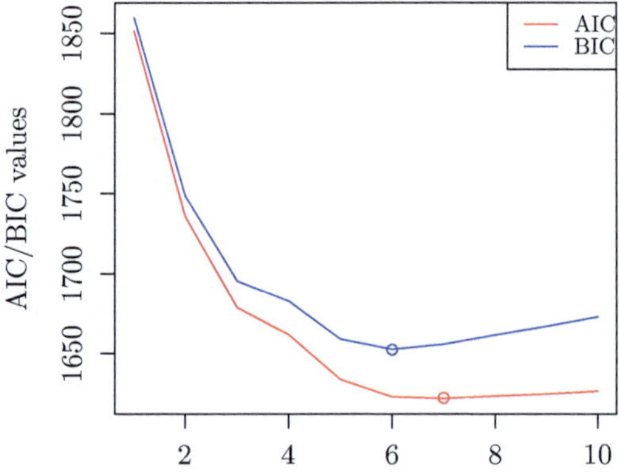

Fig. 7.2 Evaluation on the Boston dataset (Example 52). Although BIC's value itself is larger than AIC's, BIC selects a more parsimonious model (with fewer variables)

7.2 Marginal Likelihood (Discrete)

In Example 13, in the second column of Table 2.1, we have

$$\sum_{x_1=0}^{1}\sum_{x_2=0}^{1}\sum_{x_3=0}^{1} p(x_1, x_2, x_3 \mid \theta) = 1 \tag{7.7}$$

However, since $\hat{\theta}$ depends on x_1, x_2, x_3, if we write the value obtained by substituting it into the likelihood as $p(x_1, x_2, x_3 \mid \hat{\theta}(x_1, x_2, x_3))$ (the fifth column of Table 2.1), then unlike (7.7),

$$\sum_{x_1=0}^{1}\sum_{x_2=0}^{1}\sum_{x_3=0}^{1} p(x_1, x_2, x_3 \mid \hat{\theta}(x_1, x_2, x_3)) = \frac{102}{27} > 1$$

This means that maximum likelihood estimation reacts too sensitively to the data and yields estimates that are inconsistent with what would occur if other data were observed.

The above inconsistency arises because θ is unknown, and we attempt to estimate it by a single value. Therefore, consider setting values $q(x_1, x_2, x_3)$, $x_1, x_2, x_3 = 0, 1$ that satisfy

$$\sum_{x_1=0}^{1}\sum_{x_2=0}^{1}\sum_{x_3=0}^{1} q(x_1, x_2, x_3) = 1, \qquad q(x_1, x_2, x_3) \geq 0, \qquad x_1, x_2, x_3 = 0, 1 \tag{7.8}$$

For example, using a function $\varphi(\cdot)$ of θ such that

$$\int_0^1 \varphi(\theta) d\theta = 1, \qquad \varphi(\theta) \geq 0, \;\; 0 \leq \theta \leq 1 \tag{7.9}$$

we define

$$q(x_1, x_2, x_3) = \int_0^1 p(x_1, x_2, x_3 \mid \theta)\varphi(\theta) d\theta \tag{7.10}$$

As long as $\varphi(\cdot)$ satisfies (7.9), (7.8) holds. $\varphi(\cdot)$ is called a *prior distribution* for θ. Below, we consider setting the values $q(x_1, x_2, x_3)$, $x_1, x_2, x_3 = 0, 1$ using this approach. When, as in (7.10), a distribution $p(x \mid \theta)$ is defined for each parameter θ, the value

$$\int_\Theta \varphi(\theta) \prod_{i=1}^{n} p(x_i \mid \theta) d\theta$$

is called the *marginal likelihood* of the sequence $x_1, \ldots, x_n$. The marginal likelihood is also defined when a sequence of length n takes continuous values (Sect. 7.3).

Example 53 If we set $\varphi(\theta) = 1$ $(0 \le \theta \le 1)$, then

$$q(0,0,0) = q(1,1,1) = \int_0^1 \theta^3 \cdot 1 \cdot d\theta = \left[\frac{\theta^4}{4}\right]_0^1 = \frac{1}{4}$$

$$q(0,1,1) = q(1,0,0) = \int_0^1 (1-\theta)\theta^2 \cdot 1 \cdot d\theta = \left[\frac{\theta^3}{3} - \frac{\theta^4}{4}\right]_0^1 = \frac{1}{12}$$

By symmetry, the other four cases are computed similarly, and we can confirm that (7.8) holds. ■

Here, consider the case where X takes two values. As seen in Example 53, many readers may feel that it is fine to adopt a uniform prior $\varphi(\theta) = 1$. The uniform distribution appears to represent a "state of no information (uninformative)" in a natural way.

However, as the following example shows, the property of being "uniform" depends on the choice of parameterization and is only apparent.

Example 54 Assume that θ follows a uniform distribution on $[0, 1]$. Under the change of variable $t = \sqrt{\theta}$, the prior distribution for the new parameter t is

$$\frac{d\theta}{dt} = 2t, \quad \varphi(\theta)d\theta = 1 \cdot d\theta = 2t\,dt.$$

Thus, the prior for t is $\varphi(t) = 2t$, which is not uniform. ■

In this way, the uniform distribution is not invariant under reparameterization and cannot preserve "uninformativeness" with respect to coordinate transformations. Therefore, it is natural to define an "uninformative prior" that is invariant to the choice of parameterization.

From this perspective, we introduce the invariant prior proposed by Jeffreys. First, the *Fisher information matrix* for $\theta \in \Theta \subseteq \mathbb{R}^p$ is given by

$$I_{i,j}(\theta) = \int_{\mathcal{X}} \frac{\partial \log p(x \mid \theta)}{\partial \theta_i} \frac{\partial \log p(x \mid \theta)}{\partial \theta_j} p(x \mid \theta)\,dx.$$

Then the *Jeffreys prior* is defined by

$$\varphi_\Theta(\theta) = \frac{\sqrt{\det I(\theta)}}{\int_\Theta \sqrt{\det I(\theta')}\,d\theta'}.$$

Example 55 When $p = 1$, for the log-likelihood $L_\Theta := \log\{\theta^k(1-\theta)^{n-k}\}$, we have

$$I(\theta) = \mathbb{E}_X\left[\left(\frac{\partial L_\Theta}{\partial \theta}\right)^2\right] = \mathbb{E}_X\left[\left\{\frac{k - n\theta}{\theta(1-\theta)}\right\}^2\right] = \frac{n}{\theta(1-\theta)}.$$

Therefore, the Jeffreys prior is

$$\varphi_\Theta(\theta) = \frac{C_\Theta}{\sqrt{\theta(1-\theta)}}, \quad C_\Theta = \left[\int_0^1 \frac{1}{\sqrt{\theta(1-\theta)}} d\theta\right]^{-1}.$$

■

That this distribution does not depend on the choice of parameterization can be confirmed by expressing it in a different parameter and obtaining the same result.

Example 56 If we parameterize by $t = \sqrt{\theta}$ instead of θ, the log-likelihood is

$$L_T := \log\{t^{2k}(1-t^2)^{n-k}\}.$$

Then the Jeffreys prior is

$$\varphi_T(t) = \frac{C_T}{\sqrt{1-t^2}}, \quad C_T = \left[\int_0^1 \frac{1}{\sqrt{1-t^2}} dt\right]^{-1} \tag{7.11}$$

(Problem 70). If we set $C_T = 2C_\Theta$, then

$$\varphi_\Theta(\theta)\, d\theta = \frac{C_\Theta}{\sqrt{\theta(1-\theta)}} d\theta = \frac{C_\Theta}{\sqrt{t^2(1-t^2)}} \cdot \frac{d\theta}{dt} dt = \frac{C_T}{\sqrt{1-t^2}} dt = \varphi_T(t)dt$$

holds. ■

Thus, the Jeffreys prior is invariant under reparameterization and preserves a consistent "uninformativeness" regardless of representation. In this sense, Jeffreys prior can be regarded as a "truly uninformative prior."

We now consider a general case with the Beta prior $Be(a, b)$ not only for $a = b = 1$ (uniform prior) but also for $a = b = 0.5$ (Jeffreys prior), i.e., $a, b > 0$.

$$\begin{aligned} q(x_1, \dots, x_n) &:= \int_0^1 \theta^k(1-\theta)^{n-k}\varphi(\theta)d\theta = \int_0^1 \frac{\theta^{k+a-1}(1-\theta)^{n-k+b-1}}{B(a,b)} d\theta \\ &= \frac{B(k+a, n-k+b)}{B(a,b)} \end{aligned} \tag{7.12}$$

Using (2.11), we obtain

$$q(x_1, \dots, x_n) = \frac{\Gamma(k+a)\Gamma(n-k+b)\Gamma(a+b)}{\Gamma(n+a+b)\Gamma(a)\Gamma(b)} \tag{7.13}$$

In particular, for the uniform prior ($a = b = 1$),

$$q(x_1, \ldots, x_n) = \frac{k!(n-k)!}{(n+1)!} \tag{7.14}$$

holds.

Next, given observed data $x_1, \ldots, x_n$, the probabilities that $x_{n+1} = 0, 1$ are

$$q(x_{n+1} \mid x_1, \ldots, x_n) := \frac{q(x_1, \ldots, x_n, x_{n+1})}{q(x_1, \ldots, x_n)} = \begin{cases} \dfrac{k+a}{n+a+b}, & x_{n+1} = 1 \\ \dfrac{n-k+b}{n+a+b}, & x_{n+1} = 0 \end{cases} \tag{7.15}$$

(Problem 72). Indeed, from (2.8) and (7.13),

$$\begin{aligned} \frac{q(x_1, \ldots, x_n, 1)}{q(x_1, \ldots, x_n)} &= \frac{\Gamma(k+1+a)\Gamma(n-k+b)}{\Gamma(n+1+a+b)} / \frac{\Gamma(k+a)\Gamma(n-k+b)}{\Gamma(n+a+b)} \\ &= \frac{k+a}{n+a+b} \\ \frac{q(x_1, \ldots, x_n, 0)}{q(x_1, \ldots, x_n)} &= \frac{\Gamma(k+a)\Gamma(n-k+1+b)}{\Gamma(n+1+a+b)} / \frac{\Gamma(k+a)\Gamma(n-k+b)}{\Gamma(n+a+b)} \\ &= \frac{n-k+b}{n+a+b} \end{aligned}$$

Therefore, the marginal likelihood can be written as the product of these conditional probabilities. That is, with $x_i = 0, 1$,

$$\begin{aligned} q(x_1, \ldots, x_n) &= \prod_{i=1}^{n} q(x_i \mid x_1, \ldots, x_{i-1}) \\ &= \prod_{i=1}^{n} \left\{ \left(\frac{k_{i-1}+a}{i-1+a+b} \right)^{x_i} \left(\frac{i-1-k_{i-1}+b}{i-1+a+b} \right)^{1-x_i} \right\} \end{aligned}$$

where k_{i-1} denotes the number of ones among $x_1, \ldots, x_{i-1}$.

Example 57 For $n = 3$, the values of $q(x_1, x_2, x_3), x_1, x_2, x_3 = 0, 1$ are as follows: when $a = b = 0.5$,

$$q(0, 0, 0) = \frac{\frac{1}{2}}{\frac{1}{2}+\frac{1}{2}} \cdot \frac{1+\frac{1}{2}}{1+\frac{1}{2}+\frac{1}{2}} \cdot \frac{2+\frac{1}{2}}{2+\frac{1}{2}+\frac{1}{2}} = \frac{5}{16}$$

$$q(0, 0, 1) = \frac{\frac{1}{2}}{\frac{1}{2}+\frac{1}{2}} \cdot \frac{1+\frac{1}{2}}{1+\frac{1}{2}+\frac{1}{2}} \cdot \frac{\frac{1}{2}}{2+\frac{1}{2}+\frac{1}{2}} = \frac{1}{16}$$

when $a = b = 1$,

$$q(0,0,0) = \frac{1}{1+1} \cdot \frac{1+1}{1+1+1} \cdot \frac{2+1}{2+1+1} = \frac{1}{4}$$
$$q(0,0,1) = \frac{1}{1+1} \cdot \frac{1+1}{1+1+1} \cdot \frac{1}{2+1+1} = \frac{1}{12}$$

By symmetry, the remaining six values are found similarly. We see that the case $a = b = 1$ is less sensitive to the data values than the case $a = b = 0.5$. ■

If we know, for example, that the frequency of ones is large, we may set $a > b$. Since $a, b > 0$ are parameters that determine the prior φ and can be viewed as (possibly non-integer) frequencies, we will refer to them as *prior frequencies* in this book.

Furthermore, when $X_1, \ldots, X_n$ are i.i.d. and take not two values but α values in $\{1, \ldots, \alpha\}$, Proposition 24 extends as follows:

Proposition 24 *Let a random variable X take α possible values $1, \ldots, \alpha$ with frequencies $k_1, \ldots, k_\alpha$, and prior frequencies $a_1, \ldots, a_\alpha > 0$. Then the marginal likelihood of $X = x_1, \ldots, x_n$ is given by*

$$q(x_1, \ldots, x_n) = \frac{\prod_{j=1}^{\alpha} \Gamma(k_j + a_j)}{\Gamma(\sum_{j=1}^{\alpha}(k_j + a_j))} \cdot \frac{\Gamma(\sum_{j=1}^{\alpha} a_j)}{\prod_{j=1}^{\alpha} \Gamma(a_j)} \tag{7.16}$$

Proof: The marginal likelihood of whether $X = 1$ or $2 \le X \le \alpha$ is

$$\frac{B(k_1 + a_1, \sum_{h=2}^{\alpha}(k_h + a_h))}{B(a_1, \sum_{h=2}^{\alpha} a_h)} = \frac{\Gamma(k_1 + a_1)\Gamma(\sum_{h=2}^{\alpha}(k_h + a_h))}{\Gamma(\sum_{j=1}^{\alpha}(k_j + a_j))} \cdot \frac{\Gamma(\sum_{h=1}^{\alpha} a_h)}{\Gamma(a_1)\Gamma(\sum_{h=2}^{\alpha} a_h)}$$

Given $X \ge 2$, the marginal likelihood of whether $X = 2$ or $X \ge 3$ is

$$\frac{B((k_2 + a_2), \sum_{h=3}^{\alpha}(k_h + a_h))}{B(a_2, \sum_{h=3}^{\alpha} a_h)} = \frac{\Gamma(k_2 + a_2)\Gamma(\sum_{h=3}^{\alpha}(k_h + a_h))}{\Gamma(\sum_{j=2}^{\alpha}(k_h + a_h))} \cdot \frac{\Gamma(\sum_{h=2}^{\alpha} a_h)}{\Gamma(a_2)\Gamma(\sum_{h=3}^{\alpha} a_h)}$$

Multiplying these $\alpha - 1$ terms,

$$\prod_{j=2}^{\alpha} \frac{B((k_{j-1}+a_{j-1}), \sum_{h=j}^{\alpha}(k_h+a_h))}{B(a_{j-1}, \sum_{h=j}^{\alpha} a_h)} \tag{7.17}$$

gives (7.16) (Problem 75).

Using (7.17) in the proof, we write an R program to compute the marginal likelihood. Since marginal likelihoods can be too small to print conveniently, we output the negative log marginal likelihood. The number of categories is `m`, the frequencies are stored in `k[1]` through `k[m]`, and the prior frequencies in `a[1]` through `a[m]`.

```
Q.1 <- function(k,a){
  m <- length(k)
  log.q <- 0
  for(j in 2:m) {
    log.q <- log.q + log( beta(k[j-1]+a[j-1], sum(k[j:m]+a[j:m])) )
                   - log( beta(a[j-1], sum(a[j:m])))
  }
  return(-log.q)
}
```

Example run:

```
> n <- 100
> m <- 4
> prob <- rep(1/m, m)
> k <- as.vector(rmultinom(1, n, prob))
> a <- rep(1,m)
> Q.1(k,a)
[1] 143.5823
```

Additionally, when X and Y take l and m values, respectively, and we want the marginal likelihood of

$$(X, Y) = (x_1, y_1), \ldots, (x_n, y_n),$$

we may treat $Z = (X, Y)$ as a random variable taking lm values and compute accordingly. In R, the `table` function is useful for this. In the example below, `as.vector(table(x,y))` turns the four frequencies into a one-dimensional vector.

Example run:

```
> n <- 100
> x <- rbinom(n, 1, 0.5)
> y <- rbinom(n, 1, 0.25 )
> k <- as.vector(table(x,y))
```

```
> a <- rep(0.5, 4)
> Q.1(k,a)
[1] 133.8526
```

Up to now, for discrete variables, we have derived marginal likelihoods mathematically for Beta priors. However, closed-form expressions for marginal likelihoods are not available for arbitrary priors. In general, one uses *Markov chain Monte Carlo* (MCMC) to generate random draws from the posterior

$$p(\theta \mid x_1, \ldots, x_n) := \frac{\prod_{i=1}^n p(x_i \mid \theta)\varphi(\theta)}{\int_\Theta \prod_{i=1}^n p(x_i \mid \theta)\varphi(\theta)d\theta}$$

i.e., generate

$$\theta_1, \ldots, \theta_m \sim p(\cdot \mid x_1, \ldots, x_n),$$

and approximate

$$\int_\Theta \left\{\prod_{i=1}^n p(x_i \mid \theta)\right\} p(\theta \mid x_1, \ldots, x_n)d\theta \approx \frac{1}{m}\sum_{j=1}^m \left\{\prod_{i=1}^n p(x_i \mid \theta_j)\right\}.$$

In this chapter, the kinds of marginal likelihoods that admit analytical solutions are essentially limited to cases where we apply *conjugate priors* to *exponential families*.

7.3 Marginal Likelihood (Continuous)

Next, suppose we observe a sequence $x_1, \ldots, x_n \in \mathbb{R}^p$ generated from a normal distribution with unknown mean $\mu \in \mathbb{R}^p$ and covariance matrix $\Sigma \in \mathbb{R}^{p \times p}$.

Let $\kappa_0, \nu_0 > p - 1$, $\mu_0 \in \mathbb{R}^p$, and $\Lambda_0 \in \mathbb{R}^{p \times p}$, and set the prior as follows:

$$\mu \mid \Sigma \sim N(\mu_0, \Sigma/\kappa_0), \qquad \Sigma \sim W^{-1}(\Lambda_0, \nu_0) \tag{7.18}$$

Here $W^{-1}(\Lambda_0, \nu_0)$ denotes the inverse-Wishart distribution with parameters (Λ_0, ν_0), defined by (2.30). Define

$$\bar{x} := \frac{1}{n}\sum_{i=1}^n x_i, \qquad \kappa_n := \kappa_0 + n, \qquad \nu_n := \nu_0 + n$$

$$\Lambda_n := \Lambda_0 + \sum_{i=1}^n (x_i - \bar{x})(x_i - \bar{x})^T + \frac{\kappa_0 n}{\kappa_0 + n}(\bar{x} - \mu_0)(\bar{x} - \mu_0)^T$$

Then the following holds:

Proposition 25 (K. Murphy [19]) *Let $x_1, \ldots, x_n$ be a sequence generated from a normal distribution $N(\mu, \Sigma)$ with unknown mean $\mu \in \mathbb{R}^p$ and covariance $\Sigma \in \mathbb{R}^{p \times p}$. With the prior (7.18), the marginal likelihood is given by*

$$\frac{1}{\pi^{np/2}} \frac{\Gamma_p(\nu_n/2)}{\Gamma_p(\nu_0/2)} \frac{\det(\Lambda_0)^{\nu_0/2}}{\det(\Lambda_n)^{\nu_n/2}} \left(\frac{\kappa_0}{\kappa_n}\right)^{p/2}$$

Proof: See the appendix at the end of the chapter.

When we use the inverse-Wishart distribution as a prior for the covariance matrix Σ, the posterior distribution can be expressed by updating the parameters of the prior. Conversely, the Wishart distribution is conjugate to the precision matrix $\Lambda = \Sigma^{-1}$ rather than the covariance. Both are applicable, but it is often more convenient to consider a prior on the covariance matrix.

In R, the computation of $-\log p(D)$ for the marginal likelihood $p(D)$ can be implemented as follows:

```
    # Load required package
library(CholWishart)  # needed for lmvgamma

# Inputs:
#    x         : observed data (n-by-p matrix)
#    kappa.0 : prior scale parameter
#    nu.0     : prior degrees of freedom
#    mu.0     : prior mean vector
#    Lambda.0: prior covariance matrix
# Output:
#    value    : negative log marginal likelihood

Q.3 <- function(x, kappa.0, nu.0, mu.0, Lambda.0) {
  x <- as.matrix(x)
  n <- nrow(x)
  d <- ncol(x)

  kappa.n <- kappa.0 + n
  nu.n <- nu.0 + n

  x.bar <- colMeans(x)
  S <- t(x - matrix(x.bar, n, d, byrow = TRUE)) %*%
        (x - matrix(x.bar, n, d, byrow = TRUE))
  mu.diff <- matrix(x.bar - mu.0, nrow = d)
  M <- (kappa.0 * n) / (kappa.0 + n) * (mu.diff %*% t(mu.diff))
  Lambda.n <- Lambda.0 + S + M
  value <- n * d / 2 * log(pi) -
    lmvgamma(nu.n / 2, d) + lmvgamma(nu.0 / 2, d) -
    (nu.0 / 2) * log(det(Lambda.0)) + (nu.n / 2) * log(det(Lambda.n)) -
    (d / 2) * log(kappa.0 / kappa.n)
  return(value)
}
```

Example 58 We tested the above R function Q.3 on sample data.

```
> library(MASS)
> library(CholWishart)
> kappa.0 <- 1
> nu.0 <- 2
> mu.0 <- c(0, 0)
> Lambda.0 <- diag(2)
> n <- 100
> x <- mvrnorm(n, c(0,0), matrix(c(1,0,0,1), 2, 2))
> a <- rnorm(2)
> A <- a %*% t(a)
> y <- mvrnorm(n, c(0,0), A)
> Q.3(x, kappa.0, nu.0, mu.0, Lambda.0)
> Q.3(y, kappa.0, nu.0, mu.0, Lambda.0)

[1] 278.465685319517
[1] 59.697458710311
```

■

7.4 The Essence of BIC

The negative logarithm of the marginal likelihood

$$F(x_1, \dots, x_n) := -\log q(x_1, \dots, x_n) \tag{7.19}$$

is called the free energy. Except for cases where we apply conjugate priors within exponential families, an exact value (derivable in closed form) is not available, and we must resort to large-scale MCMC computations for approximation.

In this book, when the following conditions hold, we say that a statistical model $p(\cdot \mid \theta), \theta \in \Theta \subseteq \mathbb{R}^d$, is regular with respect to the true distribution $q(\cdot)$ [42].

- The Fisher information matrix has rank d.
- The minimizer $\theta = \theta_* \in \Theta$ of the Kullback–Leibler (KL) divergence from the true distribution is unique.
- θ_* does not lie on the boundary of Θ.

Here, the KL divergence is defined by

$$\int_{\mathcal{X}} q(x) \log \frac{q(x)}{p(x \mid \theta)} dx.$$

By the inequality

$$\log x \le x - 1, \qquad x > 0 \tag{7.20}$$

the KL divergence is nonnegative and equals 0 only when $p(X \mid \theta) = q(X)$ holds with probability 1. If there are multiple θ that minimize the KL divergence, the MLE is not uniquely determined.

Regularity also serves as a condition guaranteeing the existence of the MLE. Assuming regularity, it is known that the value of the free energy (7.19) can be approximated by BIC (7.2).

As a property of the Gamma function, *Stirling's formula*

$$\Gamma(x) \sim \sqrt{2\pi} e^{-x} x^{x-\frac{1}{2}} \tag{7.21}$$

is well known. Here, $f(x) \sim g(x)$ means that the ratio tends to 1 as $x \to \infty$.

Example 59 Using Stirling's formula, set $a_j = 0.5$ in (7.16); then

$$\Gamma(k_j + \tfrac{1}{2}) \sim \sqrt{2\pi} \exp\{-(k_j + \tfrac{1}{2})\} \left(k_j + \tfrac{1}{2}\right)^{k_j}$$

$$\Gamma(n + \tfrac{\alpha}{2}) \sim \sqrt{2\pi} \exp\{-(n + \tfrac{\alpha}{2})\} \left(n + \tfrac{\alpha}{2}\right)^{\alpha}$$

and hence, up to terms that converge to 0 as $n \to \infty$,

$$\sum_{j=1}^{\alpha} -k_j \log \frac{k_j + 1/2}{n + \alpha/2} + \frac{\alpha - 1}{2} \log(n+1) + \frac{\alpha - 1}{2} \log(2\pi) - \log \frac{\Gamma(\alpha/2)}{\Gamma(1/2)^{\alpha}} \tag{7.22}$$

represents $-\log q(x_1, \ldots, x_n)$ (Problem 73). Writing the deviation in constants as $O(1)$, we have

$$-\log q(x_1, \ldots, x_n) = \sum_{j=1}^{\alpha} -k_j \log \frac{k_j}{n} + \frac{\alpha - 1}{2} \log n + O(1) \tag{7.23}$$

(Problem 73). ■

Like BIC, the marginal likelihood can be used for model selection such as linear regression. If we compare the value of the marginal likelihood of a length-n sequence x^n, y^n, z^n, $Q(x^n, y^n, z^n)$, divided by the marginal likelihood of x^n, y^n, $Q(x^n, y^n)$, with the value of the marginal likelihood of x^n, z^n, $Q(x^n, z^n)$, divided by the marginal likelihood of x^n, $Q(x^n)$, we can decide whether both X and Y or only X are appropriate as explanatory variables for Z. That is,

$$\frac{Q(x^n, y^n, z^n)}{Q(x^n, y^n)} > \frac{Q(x^n, z^n)}{Q(x^n)} \iff Z \text{ uses } (X, Y) \text{as explanatory variables.}$$

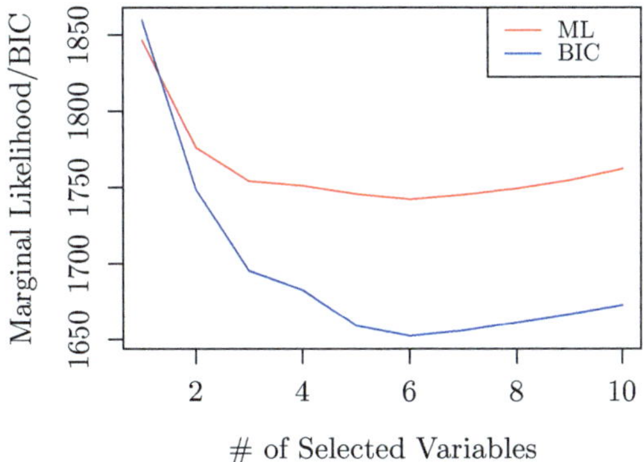

Fig. 7.3 Variation of marginal likelihood and BIC values on the Boston data

Example 60 Taking `med` (the 14th variable) in the Boston dataset as the response, to determine which variables should be used as explanatory variables, we select a subset `index` of multiple variables from among the first 13 variables that are continuous (10 variables), compute the marginal likelihood for the variables corresponding to `index` and the marginal likelihood including `med`, and choose the subset that maximizes the ratio of the two. In practice, for numerical stability, we computed the difference between the negative logs of the two. We also compared with the unsimplified BIC value `BIC.2` for continuous data. Although the values themselves differed by less than about 10%, the number of variables at which the minimum was attained agreed at 6 in both cases (Fig. 7.3).

```
# Prepare Boston data
df <- Boston
X <- as.matrix(df[, c(1, 3, 5, 6, 7, 8, 10, 11, 12, 13)]) # 10 explanatory variables
Y <- as.matrix(df[, 14]) # response variable
n <- nrow(X)
X <- X[1:n, ]
Y <- Y[1:n, ]
p <- ncol(X)

# Score computation: for each k, minimize (Q3(X,Y) - Q3(X)) over variable subsets
Q3.seq <- numeric(p)
for (k in 1:p) {
  T <- combn(1:p, k)
  m <- ncol(T)
  min_score <- Inf
  for (j in 1:m) {
    idx <- T[, j]
    X.sub <- X[, idx, drop = FALSE]
    mu.0 <- rep(0, ncol(X.sub))
    Lambda.0 <- diag(ncol(X.sub))
    q3_x <- Q.3(X.sub, 1, ncol(X.sub) + 2, mu.0, Lambda.0)
    X.sub.y <- cbind(X.sub, Y)
    mu.0.y <- rep(0, ncol(X.sub.y))
```

```
      Lambda.0.y <- diag(ncol(X.sub.y))
      q3_xy <- Q.3(X.sub.y, 1, ncol(X.sub.y) + 2, mu.0.y, Lambda.0.y)
      score <- q3_xy - q3_x
      if (score < min_score) {
        min_score <- score
      }
    }
    Q3.seq[k] <- min_score
}
# Plot
plot(1:p, Q3.seq, type = "b", col = "forestgreen", lwd = 2,
     xlab = "Number of selected variables", ylab = expression(Q[3](X,Y) -
     Q[3](X)),
     main = "Variable selection by marginal likelihood difference Q.3(X,Y)
     -Q.3(X)")
grid()
```

In general, their relationship can be described via Laplace's method. Let

$$l(\theta) := \frac{1}{n}\sum_{i=1}^{n} \log p(x_i \mid \theta), \qquad C := -\nabla^2 l(\hat{\theta}) = \left(-\frac{\partial^2 l}{\partial\theta_i \partial\theta_j}\bigg|_{\theta=\hat{\theta}} \right),$$

then the following proposition holds.

Proposition 26 (Laplace's Method) *The difference between the negative log marginal likelihood (free energy)* $-\log q(x_1, \ldots, x_n)$ *and*

$$\sum_{i=1}^{n} -\log p(x_i \mid \hat{\theta}) + \frac{d}{2}\log n - \frac{d}{2}\log 2\pi + \frac{1}{2}\log\det C - \log(\varphi(\hat{\theta}))$$

converges in probability to 0 *as* $n \to \infty$.

Proof: See, for example, [7].

As above, when the true distribution and the statistical model are regular, BIC is applicable. Although the marginal likelihood requires specifying a particular prior and large-scale computation via MCMC, the benefits are considerable.

7.5 The Essence of AIC

Although AIC and BIC may appear similar when comparing their defining formulas, they are derived from different principles.

Suppose we obtain the MLE $\hat{\theta}(x_1, \ldots, x_n)$ from observed data $x_1, \ldots, x_n$. AIC can be written as

$$AIC(x_1, \ldots, x_n) := -\sum_{i=1}^{n} \log p(x_i \mid \hat{\theta}(x_1, \ldots, x_n)) + d.$$

The second term d is derived from the relationship

$$\mathbb{E}_{X_1 \cdots X_n}[AIC(X_1, \ldots, X_n)] = \mathbb{E}_{X_1 \cdots X_n} \mathbb{E}_X [\log p(X \mid \hat{\theta}(X_1, \ldots, X_n))],$$

when taking the average over $X_1 = x_1, \ldots, X_n = x_n$. Here, $\mathbb{E}_{X_1 \cdots X_n}$ can be interpreted as the average over the training data used to obtain $\hat{\theta}$, and $\mathbb{E}_X$ as the average over test data.

Example 61 (Derivation of AIC for Linear Regression) Given observed data of explanatory and response variables $(x_1, y_1), \ldots, (x_n, y_n) \in \mathbb{R}^d \times \mathbb{R}$, we estimate β and wish to correctly predict new observations $(x_1, z_1), \ldots, (x_n, z_n) \in \mathbb{R}^d \times \mathbb{R}$. Let γ be an estimator of β, and $X = [x_1, \ldots, x_n]^\top$; we wish to minimize the average negative log-likelihood

$$\begin{aligned}
& -\log \left[\prod_{i=1}^{n} \frac{1}{\sqrt{2\pi\sigma^2}} \exp\left\{ -\frac{(z_i - x_i\gamma)^2}{2\sigma^2} \right\} \right] \\
&= \frac{n}{2} \log(2\pi\sigma^2) + \frac{\|z - X\beta + X\beta - X\gamma\|^2}{2\sigma^2} \\
&= \frac{n}{2} \log(2\pi\sigma^2) + \frac{\|z - X\beta\|^2}{2\sigma^2} - \frac{(z - X\beta)^\top X(\gamma - \beta)}{\sigma^2} + \frac{\|X(\gamma - \beta)\|^2}{2\sigma^2}
\end{aligned}$$

Averaging over $Z = z_1, \ldots, z_n$, the numerator of the second term becomes $n\sigma^2$ and the third term becomes 0.

$$\begin{aligned}
& \mathbb{E}_{Z_1 \cdots Z_n} \left\{ -\log \left[\prod_{i=1}^{n} \frac{1}{\sqrt{2\pi\sigma^2}} \exp\left\{ -\frac{(z_i - x_i\gamma)^2}{2\sigma^2} \right\} \right] \right\} \\
&= \frac{n}{2} \log(2\pi\sigma^2 e) + \frac{\|X(\gamma - \beta)\|^2}{2\sigma^2}
\end{aligned} \tag{7.24}$$

Restricting candidates for γ to unbiased estimators of β, for any unbiased estimator $\tilde{\beta}$ of β and the least-squares estimator $\hat{\beta}$, we have

$$\mathbb{E}_{Y_1 \cdots Y_n} \left[\|X(\tilde{\beta} - \beta)\|^2 \right] \geq \mathbb{E}_{Y_1 \cdots Y_n} \left[\|X(\hat{\beta} - \beta)\|^2 \right] = d\sigma^2. \tag{7.25}$$

AIC is the criterion that minimizes

$$\frac{n}{2} \log(2\pi\sigma^2 e) + \frac{d}{2}, \tag{7.26}$$

obtained by substituting the right-hand side of (7.25) into (7.24). However, $\log \sigma^2$ is unknown; if we substitute its MLE $\log \hat{\sigma}^2$, the average becomes smaller by $(d +$

$1)/n$:

$$\mathbb{E}_{Y_1\cdots Y_n}[\log\hat{\sigma}^2] = \log\sigma^2 - \frac{d+1}{n} + O\left(\frac{1}{n^2}\right). \tag{7.27}$$

Hence,

$$\mathbb{E}_{Y_1\cdots Y_n}\left[\frac{n}{2}\log(2\pi\hat{\sigma}^2 e) + \frac{d}{2} + \frac{d+1}{2} + O\left(\frac{1}{n}\right)\right]$$

agrees with (7.26) up to $O(1)$. Above, $\mathbb{E}_{Y_1\cdots Y_n}[\cdot]$ and $\mathbb{E}_{Z_1\cdots Z_n}[\cdot]$ are the averages over the training and test data, respectively. In practice, the following is often called AIC:

$$AIC := n\log(\hat{\sigma}^2) + 2d.$$

For proofs of (7.25) and (7.27), see, e.g., Chap. 5 of [34, 35]. ■

Appendix

Proof of Proposition 25

The formula for the marginal likelihood derived in this proposition coincides with (5.29) of [19], where the derivation is omitted. Here, we provide a proof by an elementary method using Bayes' theorem.

Given observations $x_1, \ldots, x_n \in \mathbb{R}^p$, the likelihood $p(D \mid \mu, \Sigma)$ can be written as

$$(2\pi)^{-\frac{np}{2}}\det(\Sigma)^{-\frac{n}{2}}\exp\left\{-\frac{1}{2}\sum_{i=1}^{n}(x_i-\mu)^\top\Sigma^{-1}(x_i-\mu)\right\} \tag{7.28}$$

On the other hand, let the prior of μ given Σ be

$$f_{\mu_0,\kappa_0}(\mu \mid \Sigma) = (\tfrac{\kappa_0}{2\pi})^{p/2}\det(\Sigma)^{-1/2}\exp\{-\tfrac{1}{2}(\mu-\mu_0)^\top(\Sigma/\kappa_0)^{-1}(\mu-\mu_0)\},$$

and take the prior of Σ to be the inverse-Wishart distribution (2.31) with $\nu = \nu_0$, $\Lambda = \Lambda_0$. Then the joint prior of (μ, Σ) is, concretely,

$$\begin{aligned} p(\mu, \Sigma) &:= f_{\mu_0,\kappa_0}(\mu \mid \Sigma) f_{\nu_0,\Lambda_0}(\Sigma) \\ &= \frac{1}{Z_0}\det(\Sigma)^{-\left(\frac{\nu_0+p}{2}+1\right)} \end{aligned}$$

$$\times \exp\left\{-\frac{1}{2}\operatorname{tr}\left(\Lambda_0\Sigma^{-1}\right)-\frac{\kappa_0}{2}(\mu-\mu_0)^\top\Sigma^{-1}(\mu-\mu_0)\right\} \tag{7.29}$$

$$Z_0=\frac{2^{\frac{\nu_0 p}{2}}\Gamma_p\left(\frac{\nu_0}{2}\right)\left(2\pi\kappa_0^{-1}\right)^{\frac{p}{2}}}{\det(\Lambda_0)^{\frac{\nu_0}{2}}}$$

We denote this prior by $NIW(\kappa_0, \mu_0, \Lambda_0, \nu_0)$, Normal-inverse Wishart. Then the product of (7.28) and (7.29), by Bayes' theorem

$$p(\mu, \Sigma \mid D)=\frac{p(\mu, \Sigma)p(D \mid \mu, \Sigma)}{p(D)},$$

is the product of the marginal likelihood $p(D)$ and the posterior $p(\mu, \Sigma \mid D)$. In particular, the exponent becomes

$$\begin{aligned}
&-\tfrac{1}{2}\operatorname{tr}(\Sigma^{-1}\Lambda_0)-\tfrac{1}{2}\sum_{i=1}^{n}(x_i-\mu)^T\Sigma^{-1}(x_i-\mu)-\tfrac{\kappa_0}{2}(\mu-\mu_0)^T\Sigma^{-1}(\mu-\mu_0)\\
&=-\tfrac{1}{2}\operatorname{tr}(\Sigma^{-1}\Lambda_0)-\tfrac{\kappa_0+n}{2}\mu^\top\Sigma^{-1}\mu+(\kappa_0\mu_0+n\bar{x})^\top\Sigma^{-1}\mu-\tfrac{\kappa_0}{2}\mu_0^\top\Sigma^{-1}\mu_0\\
&\quad-\tfrac{1}{2}\sum_{i=1}^{n}x_i^\top\Sigma^{-1}x_i\\
&=-\tfrac{1}{2}\operatorname{tr}(\Sigma^{-1}\Lambda_0)-\tfrac{\kappa_0+n}{2}\left(\mu-\frac{\kappa_0\mu_0+n\bar{x}}{\kappa_0+n}\right)^\top\Sigma^{-1}\left(\mu-\frac{\kappa_0\mu_0+n\bar{x}}{\kappa_0+n}\right)\\
&\quad-\frac{n\kappa_0}{2(\kappa_0+n)}(\mu_0-\bar{x})^\top\Sigma^{-1}(\mu_0-\bar{x})-\tfrac{1}{2}\sum_{i=1}^{n}(x_i-\bar{x})^\top\Sigma^{-1}(x_i-\bar{x})\\
&=-\tfrac{1}{2}\operatorname{tr}\left(\Sigma^{-1}\left\{\sum_{i=1}^{n}(x_i-\bar{x})(x_i-\bar{x})^T+\Lambda_0+\frac{n\kappa_0}{2(\kappa_0+n)}(\mu_0-\bar{x})(\mu_0-\bar{x})^\top\right\}\right)\\
&\quad-\tfrac{\kappa_0+n}{2}\left(\mu-\frac{\kappa_0\mu_0+n\bar{x}}{\kappa_0+n}\right)^\top\Sigma^{-1}\left(\mu-\frac{\kappa_0\mu_0+n\bar{x}}{\kappa_0+n}\right)\\
&=-\tfrac{1}{2}\operatorname{tr}(\Sigma^{-1}\Lambda_n)-\tfrac{\kappa_n}{2}(\mu-\mu_n)^\top\Sigma^{-1}(\mu-\mu_n)
\end{aligned}$$

where we set $\mu_n := \dfrac{\kappa_0\mu_0+n\bar{x}}{\kappa_0+n}$ (Problem 78), and for the third equality, we used $\operatorname{tr}(AB)=\operatorname{tr}(BA)$ whenever both products are defined (Problem 41).

For the non-exponential parts, we have

$$(2\pi)^{-\frac{np}{2}}\det(\Sigma)^{-\frac{n}{2}}\cdot\frac{1}{Z_0}\det(\Sigma)^{-\left(\frac{\nu_0+p}{2}+1\right)}=(2\pi)^{-\frac{np}{2}}\frac{Z_n}{Z_0}\cdot\frac{1}{Z_n}\det(\Sigma)^{-\left(\frac{\nu_n+p}{2}+1\right)},$$

with

$$Z_n = \frac{2^{\frac{\nu_n p}{2}} \Gamma_p\left(\frac{\nu_n}{2}\right)\left(2\pi\kappa_n^{-1}\right)^{\frac{p}{2}}}{\det(\Lambda_n)^{\frac{\nu_n}{2}}}.$$

Hence, overall,

$$(2\pi)^{-\frac{np}{2}} \frac{Z_n}{Z_0} \cdot \frac{1}{Z_n} \det(\Sigma)^{-\left(\frac{\nu_n+p}{2}+1\right)}$$
$$\times \exp\left\{-\frac{1}{2}\operatorname{tr}(\Sigma^{-1}\Lambda_n) - \frac{\kappa_n}{2}(\mu-\mu_n)^\top\Sigma^{-1}(\mu-\mu_n)\right\}$$

The part after "·" is the density of $NIW(\kappa_n, \mu_n, \Lambda_n, \nu_n)$, so integrating over (μ, Σ) yields 1 and corresponds to the posterior $p(\mu, \Sigma \mid D)$. Therefore, $(2\pi)^{-\frac{np}{2}}\dfrac{Z_n}{Z_0}$ corresponds to the marginal likelihood $p(D)$. Computing this gives

$$\frac{2^{\nu_n p/2}\Gamma_p(\nu_n/2)(2\pi/\kappa_n)^{p/2}}{\det(\Lambda_n)^{\nu_n/2}} \frac{\det(\Lambda_0)^{\nu_0/2}}{2^{\nu_0 p/2}\Gamma_p(\nu_0/2)(2\pi/\kappa_0)^{p/2}} \frac{1}{(2\pi)^{np/2}}$$
$$= \frac{1}{(2\pi)^{np/2}} \frac{2^{\nu_n p/2}}{2^{\nu_0 p/2}} \frac{(2\pi/\kappa_n)^{p/2}}{(2\pi/\kappa_0)^{p/2}} \frac{\Gamma_p(\nu_n/2)}{\Gamma_p(\nu_0/2)} \frac{\det(\Lambda_0)^{\nu_0/2}}{\det(\Lambda_n)^{\nu_n/2}},$$

which proves the proposition. ■

Problems 70-78

70. Show Eq. (7.11).
71. Verify that in Example 57, the calculation of the marginal likelihood when $a = b = 1$ coincides with the result obtained in (7.14).
72. Write an R program to generate 100 binary sequences of length 50 (where the probability of a "1" is p). For each sequence, plot $q(1 \mid x_1, \ldots, x_n)$ $(0 \le n \le 49)$ defined in (7.15), connecting the points with line segments. Perform this for $p = 0.25$, $p = 0.5$, and for $a = b = 0.5$, $a = b = 1$, obtaining a total of four graphs. Discuss what characteristics can be observed among these cases.
73. Show (7.22). Furthermore, use the inequalities

$$0 \le \sum_{j=1}^{\alpha} k_j \log(k_j+\tfrac{1}{2}) - \sum_{j=1}^{\alpha} k_j \log k_j = \sum_{j=1}^{\alpha} k_j \log(1+\tfrac{1}{2k_j}) \le \sum_{j=1}^{\alpha} k_j \cdot \tfrac{1}{2k_j} = \tfrac{\alpha}{2}$$

$$0 \le \sum_{j=1}^{\alpha} k_j \log(n + \tfrac{\alpha}{2}) - \sum_{j=1}^{\alpha} k_j \log n = \sum_{j=1}^{\alpha} k_j \log(1 + \tfrac{\alpha}{2n}) \le \sum_{j=1}^{\alpha} k_j \cdot \tfrac{\alpha}{2n} = \tfrac{\alpha}{2}$$

to show (7.23).

74. The following is a procedure to compute the minimum of the negative log marginal likelihood. Use this to modify the program in Example 52 and draw a graph similar to Fig. 7.3.

```
for (k in 1:p) {
  Q.min <- Inf
  for (j in 1:m) {
    q <- c(1, T[, j] + 1)
    S <- sigma2(X[, q], y)
    Q <- Q.3(X[, q], y, rep(0, k + 1), diag(k + 1), 1, 0.5)
    if (Q < Q.min) Q.min <- Q
  }
  Q.seq <- c(Q.seq, Q.min)
}
```

75. Derive (7.16) from (7.17).

76. Stirling's formula: with $C = \sqrt{2\pi}$, it is known that

$$Cx^{x-\frac{1}{2}}e^{-x} \le \Gamma(x) \le Cx^{x-\frac{1}{2}}e^{-x+\frac{1}{12x}}.$$

Define three functions in your program and verify this inequality.

```
C <- sqrt(2 * pi)
f <- function(x) # lower bound
g <- function(x) # true log Gamma(x)
h <- function(x) # upper bound
curve(g(x)-f(x), xlim=c(3,5), ylim = c(0, 0.03),
      xlab = "x", ylab = "log Gamma(x)",
      main = "Stirling approximation vs log Gamma(x) (zoomed)",
      col = "red", lwd = 2)

curve(h(x)-g(x), add = TRUE, col = "blue", lwd = 2, lty = 2)
abline(h=0)
```

77. Construct a procedure to compute the AIC. Fill in the blank, run the procedure, and then similarly construct and run a procedure for BIC.

```
AIC.value <- function(X,y){
    n <- length(y)
    X <- as.matrix(X)
    k <- ncol(X)
    n <- length(y)
    if(k==0){
      y.bar <- mean(y)
      S <- sum((y-y.bar)^2)/n}
    else     S <- sum((lm(y~X)$fitted.values-y)^2) / n
    return(## fill in here ##)
}
```

78. In the proof of Proposition 25, show the second equality in the derivation of the exponent using the following identities:

$$\begin{cases} a\mu^\top\Sigma^{-1}\mu + b^\top\Sigma^{-1}\mu = a(\mu+\frac{b}{2a})^\top\Sigma^{-1}(\mu+\frac{b}{2a}) - \frac{1}{4a}b^\top\Sigma^{-1}b \\ a = -\frac{\kappa_0+n}{2}, \quad b = n\bar{x}+\kappa_0\mu_0 \\ -\frac{1}{4a}b^\top\Sigma^{-1}b = \frac{1}{2(\kappa_0+n)}(n\bar{x}+\kappa_0\mu_0)^\top\Sigma^{-1}(n\bar{x}+\kappa_0\mu_0) \end{cases}$$

$$\begin{cases} \frac{1}{2(\kappa_0+n)}(n\bar{x}+\kappa_0\mu_0)^\top\Sigma^{-1}(n\bar{x}+\kappa_0\mu_0) - \frac{\kappa_0}{2}\mu_0^\top\Sigma^{-1}\mu_0 \\ = a_0\mu_0^\top\Sigma^{-1}\mu_0 + b_0^\top\Sigma^{-1}\mu_0 + c_0 = a_0(\mu_0+\frac{b_0}{2a_0})^\top\Sigma^{-1}(\mu_0+\frac{b_0}{2a_0}) \\ \qquad +c_0 - \frac{1}{4a_0}b_0^\top\Sigma^{-1}b_0 \\ a_0 = -\frac{n\kappa_0}{2(\kappa_0+n)}, \quad b_0 = \frac{n\kappa_0}{\kappa_0+n}\bar{x}, \quad c_0 = \frac{n^2}{2(\kappa_0+n)}\bar{x}^\top\Sigma^{-1}\bar{x} \\ c_0 - \frac{1}{4a_0}b_0^\top\Sigma^{-1}b_0 = \frac{n}{2}\bar{x}^\top\Sigma^{-1}\bar{x} \end{cases}$$

$$\frac{n}{2}\bar{x}^\top\Sigma^{-1}\bar{x} - \frac{1}{2}\sum_{i=1}^n x_i^\top\Sigma^{-1}x_i = -\frac{1}{2}\sum_{i=1}^n (x_i-\bar{x})^T\,\Sigma^{-1}\,(x_i-\bar{x})$$

Chapter 8
Score-Based Structure Learning

First, we introduce structure search via hill climbing using the R package `bnlearn`. With the Alarm dataset, we visualize how graphs are constructed based on scores (such as Bayesian Information Criterion (BIC)).

Next, we present a method for sequentially determining each variable's parent set using information criteria (Akaike Information Criterion (AIC) and BIC) when the variable ordering is known. We explain that a graph constructed within this framework becomes a Bayesian network by using the notion of a boundary directed acyclic graph (DAG), and we also provide a proof.

In the latter half, we introduce scores based on marginal likelihood and discuss their applicability to independence and conditional independence testing. In particular, through a comparison between scores using Jeffreys' prior and the Bayesian Dirichlet equivalent uniform (BDeu) score, we detail a practical caveat: the lack of regularity in BDeu.

Finally, we present a forest-structure learning method that extends the Chow–Liu algorithm, and we show that by introducing a new estimator in place of the conventional mutual information estimator, it may be possible to improve the accuracy of independence testing.

8.1 Overview of Score-Based Structure Learning

From n samples, we consider the problem of learning the structure (edge set) of a graphical model $G = (V, E)$ that represents dependencies among p random variables (structure learning). Let $\mathcal{X}_j$ denote the set of possible values (e.g., the real line, a finite set) for random variable X_j, and suppose the data are represented as

$$x_i = (x_i^{(1)}, \dots, x_i^{(p)}) \in \mathcal{X}_1 \times \cdots \mathcal{X}_p, \qquad i = 1, \dots, n$$

or by a matrix $X = (x_i^{(j)}) \in (\mathcal{X}_1 \times \cdots \mathcal{X}_p)^n$.

J. Suzuki, *Graphical Models and Causal Discovery with R*,
https://doi.org/10.1007/978-981-95-4267-3_8

Graphical models simultaneously represent multiple conditional independencies among random variables. Therefore, one approach is to select a structure that encodes those conditional independencies that were deemed true by statistical tests. This method is called *constraint-based* structure learning. Algorithms such as the Peter–Clark (PC) algorithm considered in Chap. 5 fall into this category.

In this chapter, we examine methods that compute a score for each candidate structure using information criteria and marginal likelihood, and then select the structure with the best score. The outcome of structure learning will differ depending on which score (AIC, BIC, or marginal likelihood) is used. Whereas the PC algorithm assumed faithfulness, *score-based* structure learning does not impose such an assumption, and therefore finding the optimal solution can require huge computational effort. Consequently, it is realistic to seek approximate solutions.

Among Bayesian network (BN) packages available on CRAN, Marco Scutari's `bnlearn` [22] is well known. The Alarm dataset (discrete, $p = 37$) is also a popular benchmark for BN structure learning. To grasp the overall picture of this chapter, we first confirm the behavior of the *hill climbing* procedure in `bnlearn` for obtaining approximate solutions. In BN structure learning, the input is always an $n \times p$ data frame, and the output is a BN structure.

Here, hill climbing starts from an initial state (an initial BN structure) and iteratively applies the operation (adding, removing, or reversing an edge) that yields the greatest score improvement. When no further improvement is possible, the algorithm stops and the current structure is regarded as optimal.

Such methods can obtain a structure relatively quickly, but because a locally optimal move is chosen at each step, a globally optimal solution may not be reached. Moreover, the obtained solution depends on the initial state. Unless the true structure is known, we cannot tell how close the hill-climbing solution is to the optimum. Hence, it is common to run the procedure multiple times with random initializations.

```
library(bnlearn)
library(igraph)
# Load the alarm data
df <- alarm
p <- ncol(df)
# Rename nodes as "1", "2", ..., "p" (for bnlearn)
colnames(df) <- as.character(1:p)
# Structure learning (Hill Climbing + BIC)
alarm.hc <- hc(df, score = "bic")
# Get (from, to) edge info as strings
mat <- arcs(alarm.hc)
# Convert strings to integers (for igraph)
u <- as.integer(mat[,1])  # parent node
v <- as.integer(mat[,2])  # child node
edges <- cbind(u, v)      # edge matrix
```

The resulting `edges` is a matrix (number of edges $\times$ 2) whose rows are ordered pairs of edge endpoints. To visualize it, we typically use the graph-drawing package `igraph`. However, when the number of nodes or edges is large, drawings can easily become cluttered. In Fig. 8.1, the figure is drawn using Cytoscape—

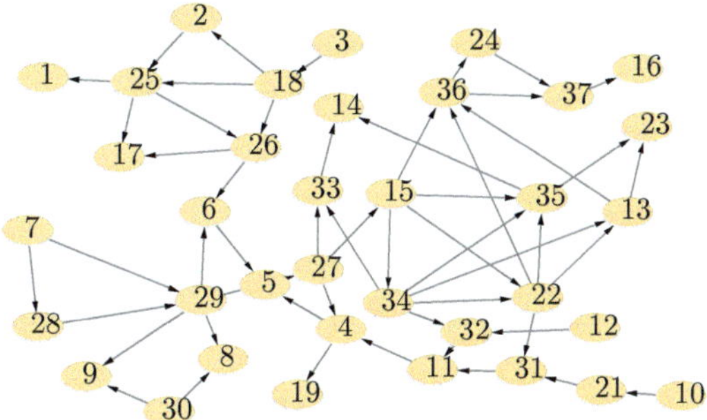

Fig. 8.1 BN learned by hill climbing in `bnlearn` on the Alarm dataset with $n = 37$. Drawn using Cytoscape

software specialized for graph visualization—based on the edge information. For presentations or publications, and especially for visualizing large networks, such specialized visualization tools are often used instead of `igraph`.

```
write.csv(edges, file = "alarm.csv", row.names = FALSE)
```

The resulting "alarm.csv" is then imported into Cytoscape.

We construct a BN according to a certain ordering; that is, there is a fixed order $1 < 2 < \cdots < p$, and if a directed edge in the DAG goes from X_i to X_j, then it must satisfy $i < j$. Thus, when the set of vertices is $V = \{1, \ldots, p\}$, for each $j = 1, \ldots, p$ we select $\pi_j \subseteq \{1, \ldots, j-1\}$, and thereby the directed edge set

$$E = \cup_{k=1}^{p} \{(j, k) \mid j \in \pi_k\}$$

is determined. In this case, we call $X_{\pi_k} = (X_j)_{j \in \pi_k}$ the *parent set* of X_k. Then

$$X_k \perp\!\!\!\perp X_{\{1,\ldots,k-1\}-\pi_k} \mid X_{\pi_k}$$

holds, and for any $\pi \subsetneq \pi_k$,

$$X_k \perp\!\!\!\perp X_{\{1,\ldots,k-1\}-\pi} \mid X_{\pi}$$

does not hold. For each $k = 1, \ldots, p$, we select such a π_k. Below, the graph $G = (V, E)$ thus constructed (under the fixed ordering) is called the *boundary DAG* for that ordering. In score-based structure learning, one first fixes the variable order and then selects such a parent set π_k for each node k. Here, we show why the boundary DAG yields a BN and provide a proof.

Proposition 27 (Verma–Pearl [20, 40]) *Given an arbitrary ordering of the random variables, the corresponding boundary DAG is a BN.*

Proof See the Appendix.

Proposition 27 is one of the most important results in score-based structure learning.

In this chapter, we examine structure learning for both cases: when the variable order is known and when it is unknown. If the variable order is known, then for each node $k = 2, \ldots, p$, we simply identify the parent set from among the upstream nodes:

$$\pi_1 = \{\}, \quad \pi_2 \subseteq \{1\}, \quad \ldots, \quad \pi_k \subseteq \{1, \ldots, k-1\}, \quad \cdots, \quad \pi_p \subseteq \{1, \ldots, p-1\}.$$

In the general setting where the variable order is unknown, we nonetheless start from the results for the known-order case.

Even when the order is unknown, structure learning proceeds by optimization—maximizing the posterior probability or minimizing an information criterion. Alternatively, although computationally more demanding, one can evaluate scores across different orderings and choose the structure with the optimal value. (We discuss the latter in the latter part of Sect. 8.3.) However, even in this approach, one cannot determine superiority among Markov equivalent structures. Accordingly, one sometimes first determines the order using LiNGAM introduced in Chap. 6 and then selects the elements of each parent set from among upstream variables.

8.2 Structure Learning via Information Criteria

In this section, assuming that the variable order is known, we describe a method to find parent sets using the information criteria AIC and BIC [30].

When all random variables are discrete, the parameters are the conditional probabilities of $X_k = x^{(k)}$ given the configuration (state) of parent variables $X_{\pi_k} = (x^{(j)})_{j \in \pi_k}$. If X_k takes α_k distinct values (with $\alpha_k - 1$ free parameters due to the sum-to-one constraint) and the number of states $s = (x^{(j)})_{j \in \pi_k}$ is $\prod_{j \in \pi_k} \alpha_j$, then the number of parameters is (Fig. 8.2)

$$d_k = (\alpha_k - 1) \prod_{j \in \pi_k} \alpha_j .$$

Suppose we have n observations of $(X_1, \ldots, X_p)$:

$$(x_i^{(1)}, \ldots, x_i^{(p)})_{i=1}^n .$$

Let n_s be the frequency of state s, and let $n_{h,s}$ be the joint frequency of state s and value h. Then the likelihood is

$$\prod_s \prod_{h=1}^{\alpha_k} \left(\frac{n_{h,s}}{n_s} \right)^{n_{h,s}} ,$$

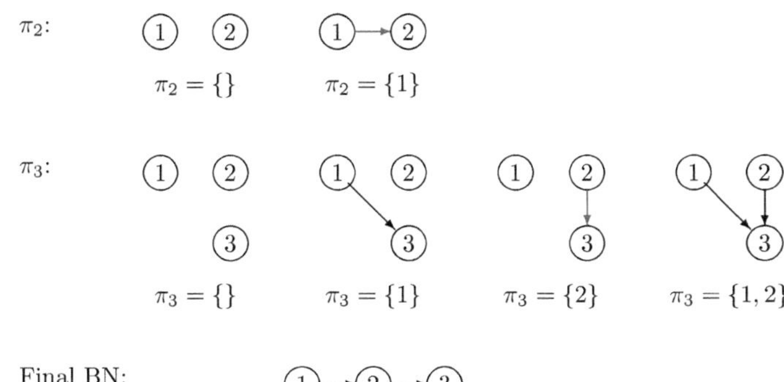

Fig. 8.2 From $\pi_1 = \{\}$, $\pi_2 = \{1\}$, and $\pi_3 = \{2\}$, the final BN $1 \to 2 \to 3$ is obtained

where $\sum_s n_s = n$ and $\sum_{h=1}^{\alpha_k} n_{h,s} = n_s$ hold. The negative log-likelihood (the **empirical entropy**) is

$$H_k = -\sum_s \sum_{h=1}^{\alpha_k} n_{h,s} \log \frac{n_{h,s}}{n_s},$$

so AIC and BIC are, respectively,

$$AIC := H_k + d_k \tag{8.1}$$

$$BIC := H_k + \frac{d_k}{2} \log n, \tag{8.2}$$

and the parent set π_k is estimated as the one that minimizes these quantities. This procedure is carried out for $k = 2, 3, \ldots, p$ (with $\pi_1 = \{\}$).

Example 62 Let $p = 3$, $n = 10$, $\alpha_1 = 2$, $\alpha_2 = 2$, $\alpha_3 = 3$, and suppose the following data are given:

```
X.1 <- c(2, 2, 2, 2, 1, 2, 2, 2, 1, 2)
X.2 <- c(2, 2, 1, 1, 2, 1, 1, 2, 2, 2)
X.3 <- c(1, 1, 2, 2, 3, 1, 3, 1, 1, 1)
```

Using the function `ftable`,

```
tab.1 <- ftable(X.1)
tab.2 <- ftable(X.2)
tab.3 <- ftable(X.3)
tab.1.2 <- ftable(X.1, X.2)
tab.1.3 <- ftable(X.1, X.3)
tab.2.3 <- ftable(X.2, X.3)
tab.1.2.3 <- ftable(X.1, X.2, X.3)
```

```
> tab.1
X.1 1 2
    2 8
>
> tab.2
X.2 1 2
    4 6
>
> tab.3
X.3 1 2 3
    6 2 2
>
>
```

```
> tab.1.2
    X.2 1 2
X.1
1       0 2
2       4 4
>
> tab.1.3
    X.3 1 2 3
X.1
1       1 0 1
2       5 2 1
>
>
```

```
> tab.2.3
    X.3 1 2 3
X.2
1       1 2 1
2       5 0 1
>
> tab.1.2.3
        X.3 1 2 3
X.1 X.2
1   1       0 0 0
    2       1 0 1
2   1       1 2 1
    2       4 0 0
```

Fig. 8.3 Cross-tabulation table obtained by executing the R function `ftable`

we obtain the contingency tables (Problem 80). The results are shown in Fig. 8.3.

Their empirical entropies are given by

```
> c(h(tab.1), h(tab.2), h(tab.3), h(tab.1.2), h(tab.1.3), h(tab.2.3),
h(tab.1.2.3))
[1] 5.004024 6.730117 9.502705 5.545177 8.588343 6.862250 5.545177
```

Here, h computes the empirical entropy from a contingency table and is defined as follows:

```
h <- function(tab) {
   m <- nrow(tab)    # number of rows (e.g., number of values of
   conditioning variable X)
   r <- ncol(tab)    # number of columns (e.g., number of values for
   response Y)
   h <- 0            # initialize entropy
   for(i in 1:m) {
      ss <- sum(tab[i, ])  # row sum (proportional to P(X = i))
      if(ss > 0) {
         for(j in 1:r) {
            if(tab[i, j] > 0) {
               # term for conditional probability p(Y = j | X = i)
               h <- h - tab[i, j] * log(tab[i, j] / ss)
            }
         }
      }
   }
   return(h)
}
```

Since $0.5 \log n = 1.15$, for π_2 and π_3 we have, respectively,

π_2	H_2	d_2	AIC	BIC
{}	6.73	1	7.73	7.88
{1}	5.55	2	7.54	7.84

π_3	H_3	d_3	AIC	BIC
{}	9.50	2	11.50	11.80
{1}	8.59	4	12.59	13.19
{2}	6.86	4	10.86	11.46
{1, 2}	5.55	8	13.55	14.75

Thus, both AIC and BIC select $\pi_2 = \{1\}$ and $\pi_3 = \{2\}$. As a result, we can estimate the BN structure (Fig. 8.2). ■

Functions to compute AIC and BIC are as follows:

```
IC <- function(X, y, proc= "AIC"){
    n <- nrow(X)
    tab <- ftable(data.frame(X,y))
    m <- nrow(tab)
    r <- ncol(tab)
    d <- m*(r-1)
    if(proc=="AIC")     IC <- h(tab)+d else
        IC <- h(tab)+d*log(n)/2
    return(IC)
}
```

The same applies to continuous variables. Using the functions `AIC.3` or `BIC.3` constructed in Chap. 6, we can obtain the parent set π_k for each $k = 2, \ldots, p$. Below is a function that, for all subsets of explanatory variables, computes AIC or BIC and selects the subset with the minimum information criterion (IC).

```
IC.min <- function(X, y, proc = "AIC") {
    # X: matrix of explanatory variables (n x p)
    # y: response vector (length n)
    # proc: "AIC" or "BIC"
    p <- ncol(X)
    n <- length(y)
    L <- 2^p
    # IC for the model with no explanatory variables (intercept-only model
    )
    IC.min <- n * log(sum((y - mean(y))^2) / n)
    set.min <- NULL
    if (p == 0) {
        return(list(value = IC.min, set = set.min))
    }
    for (i in 2:L) {
        set <- decimal.to.set(i - 1)
        if (proc == "AIC") {
            value <- AIC.3(as.matrix(X[, set]), y)
        } else {
            value <- BIC.3(as.matrix(X[, set]), y)
        }
        if (value < IC.min) {
            IC.min <- value
            set.min <- set
```

```
        }
    }
    return(list(value = IC.min, set = set.min))
```

Here, the functions AIC.3 and BIC.3 are those defined in Chap. 7.

Example:

```
> library(MASS)
> df <- Boston
> index <- c(1,3,5,6,7,8,10,11,12,13)
> X <- as.matrix(df[, index])
> y <- df[[14]]
>
> IC.min(X, y, "AIC")
$value
1655.20347628248
$set
1, 3, 4, 6, 8, 9, 10
>
> IC.min(X, y, "BIC")
$value
1679.922
$set
4, 6, 8, 9, 10
```

The function decimal.to.set converts a nonnegative integer into the subset of indices corresponding to 1-bits in its binary expansion:

```
decimal.to.set <- function(x) if(x==0) NULL else
    c(decimal.to.set(x-2^floor(log2(x))), floor(log2(x))+1)
```

This maps any integer between 0 and $2^L - 1$ to a subset of $\{1, \ldots, L\}$.

By determining the parent sets in order, we can construct the BN:

```
p <- ncol(X)
parent <- list()
parent[[1]] <- NULL
for(j in 2:p)parent[[j]] <- IC.min(as.matrix(X[, 1:(j-1)]), X[,j], "BIC")$
    set
parent
```

At this point, the BN is mathematically determined. To visualize it, we use igraph.

```
# Library for graph plotting
library(igraph)
# Pre-allocate a matrix for directed edges (up to 1000 edges)
edge <- matrix(ncol = 2, nrow = 1000)
# Edge counter
p <- 10
k <- 0
# Add a directed edge j -> i for each parent j of variable i
for (i in 1:p) {
  for (j in parent[[i]]) {
    k <- k + 1
    edge[k, ] <- c(j, i)  # register directed edge j -> i
```

```
  }
}
# Create igraph object from edge list
g <- graph_from_edgelist(edge[1:k, ], directed = TRUE)
# Label each node (variable)
# Labels are the initial letters of variable names (e.g., crim -> "C",
    indus -> "I", ...)
V(g)$label <- c("C", "I", "N", "R", "A", "R", "T", "P", "B", "L")
# Plot with default style
layout <- layout_with_fr(g)
plot(g,
     layout = layout,
     vertex.label.cex = 1,
     vertex.size = 30,
     vertex.label.color = "black",
     vertex.color = "yellow",
     edge.arrow.size = 0.5
)
```

The resulting graphs are shown in Fig. 8.4. We see that the BN obtained with AIC has more edges than the BN obtained with BIC.

8.3 Structure Learning via Marginal Likelihood

In this section, we examine the problem of finding the BN structure that maximizes the marginal likelihood.

We begin with the case of $p = 2$ random variables. There are two undirected graphs (independent or not) and three DAGs (Fig. 3.1). Suppose we observe samples of random variables X, Y: $x^n := (x_1, \ldots, x_n)$, $y^n := (y_1, \ldots, y_n)$.

For the moment, assume the random variables are discrete. First, using an unknown parameter θ_{XY} for the joint distribution of $(X, Y) = (x_i, y_i), i = 1, \ldots, n$, write $P(X = x_i, Y = y_i \mid \theta_{XY})$. Next, by marginalizing each of X and Y, write the marginal distributions as $P(X = x_i \mid \theta)$, $P(Y = y_i \mid \theta)$. We wish to infer whether $P(X, Y) = P(X)P(Y)$ (i.e., X and Y are independent) holds. However, it is difficult to reach a conclusion from this information alone. We therefore assume the prior probability of independence is $0 < p < 1$, and that the prior distributions

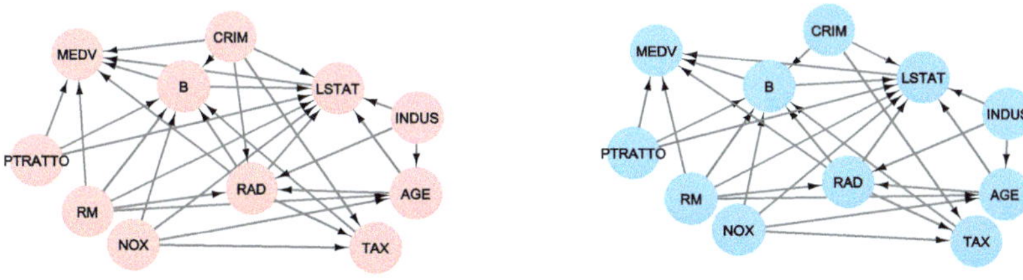

Fig. 8.4 BNs obtained from the Boston dataset using AIC (left) and BIC (right)

for θ_X, θ_Y, and θ_{XY} are π_X, π_Y, and π_{XY}, respectively. Suppose we can compute, by some method,

$$Q_X(x^n) := \int_{\Theta_X} \prod_{i=1}^{n} P(X = x_i \mid \theta)\pi_X(\theta)\, d\theta$$

$$Q_Y(y^n) := \int_{\Theta_Y} \prod_{i=1}^{n} P(Y = y_i \mid \theta)\pi_Y(\theta)\, d\theta$$

$$Q_{XY}(x^n, y^n) := \int_{\Theta_X \times \Theta_Y} \prod_{i=1}^{n} P(X = x_i, Y = y_i \mid \theta)\pi_{XY}(\theta)\, d\theta$$

where Θ_X, Θ_Y are the ranges of θ_X, θ_Y, respectively.

Then consider the decision rule that declares X and Y independent if

$$p\, Q_X(x^n)Q_Y(y^n) \geq (1-p)\, Q_{XY}(x^n, y^n), \tag{8.3}$$

and dependent otherwise. In this case, we are making the decision that maximizes the posterior probability under the assumed priors. The conclusion depends on the choice of the prior parameters p, π_X, π_Y, π_{XY}. However, it is known that as the sample size n becomes large, the influence of such priors becomes negligible.

Example 63 We implement a procedure to determine whether length-n binary sequences for X and Y are independent. The function `Q.1` is the same as that created in Chap. 7.

```
# Data generation
n <- 100 # sample size
# x is drawn from a binomial with P(x=1)=1/4
x <- rbinom(n, 1, 1/4)
k.x <- as.vector(table(x)) # counts by x category
# y is drawn from a binomial with P(y=1)=1/2
y <- rbinom(n, 1, 1/2)
k.y <- as.vector(table(y)) # counts by y category
# z encodes the (x,y) pairs into 4 categories
z <- x + 2 * y
k.z <- as.vector(table(z)) # counts by (x,y) category
# Independence decision
# If x and y are independent, the result should be negative
result <- Q.1(k.x, rep(1/2, 2)) + Q.1(k.y, rep(1/2, 2)) - Q.1(k.z, rep(1/2, 4))
# Output the result
print(result)
```

We also experimented by changing the data-generating process as follows:

```
x <- rbinom(n, 1, 1/2)
k.x <- as.vector(table(x))
```

```
y <- (rbinom(n, 1, 1/3) + x) %% 2
k.y <- as.vector(table(y))
z <- x + 2 * y
k.z <- as.vector(table(z))
```

Since `Q.1` is the negative log marginal likelihood,

```
Q.1(k.x, rep(1/2, 2)) + Q.1(k.y, rep(1/2, 2))
      - Q.1(k.z, rep(1/2, 4))
```

being negative is taken to indicate independence. The results of the decisions (500 samples for each setting) are shown in Fig. 8.5a,b. In both cases, the samples plotted above the red line were judged independent. Here we set $p = 0.5$ and the prior frequency to 0.5 (Jeffreys' prior) in (8.3).

■

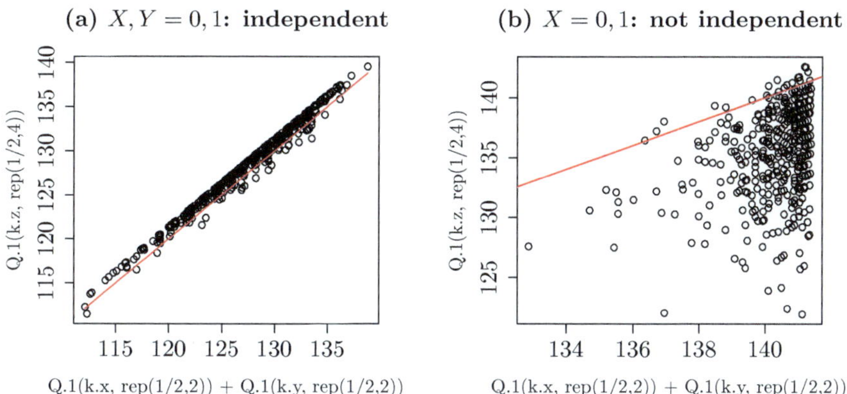

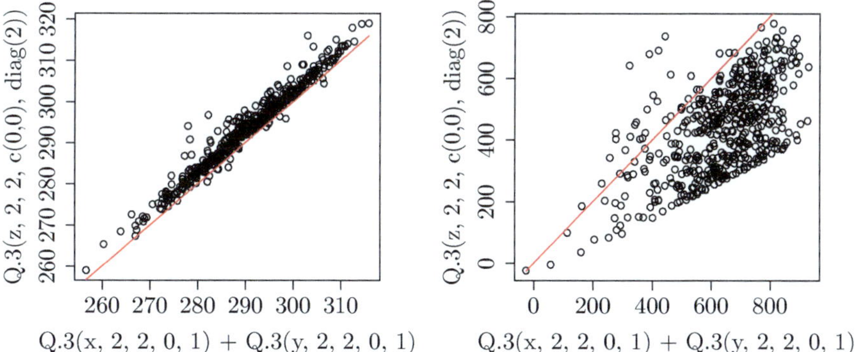

Fig. 8.5 Independence decisions based on (8.3) (Example 63). Roughly, the independent data in (**a**) and (**c**) lie above the red line, while the dependent data in (**b**) and (**d**) lie below the red line

In general, it is known that following this procedure yields the correct decision about whether X and Y are independent. The property that the probability of a correct decision converges to 1 as the sample size n grows is called **consistency**. We provide a proof for the binary case.

Proposition 28 *Let X and Y be random variables taking values in $\{1, \dots, \alpha\}$ and $\{1, \dots, \beta\}$, respectively (each following a multinomial model). Let x^n, y^n be length-n independent realizations of X and Y, respectively. Then, with appropriate parameter choices,*

$$-\log Q_X(x^n) - \log Q_Y(y^n) + \log Q_{XY}(x^n, y^n)$$

being nonpositive implies independence of X and Y, and being positive implies dependence; moreover, the probability that this decision is correct converges to 1 as n increases.

Proof First, suppose $k = (k_1, \dots, k_\alpha)$ follows the multinomial distribution with probabilities $p = (p_1, \dots, p_\alpha)$. Define

$$h(k, p) := \sum_{j=1}^{\alpha} -k_j \log p_j, \qquad \hat{p} = \left(\frac{k_1}{n}, \dots, \frac{k_\alpha}{n}\right).$$

We show that

$$D(k, p) := h(k, \hat{p}) - h(k, p) = O_P(1), \tag{8.4}$$

where $O_P(1)$ denotes a term that converges in distribution to a nondegenerate random variable as $n \to \infty$.

By the mean value form of Taylor's theorem,

$$\log(1 + z) = z - \frac{z^2}{2(1 + \xi)^2}, \qquad \xi \text{ lies between 0 and } z.$$

Applying this with $\hat{p}_j = k_j/n$, $\delta_j = \hat{p}_j - p_j$, and $z = \delta_j/p_j$, we obtain

$$\begin{aligned}
\mathrm{D}(k, p) &:= -n \sum_{j=1}^{\alpha} \hat{p}_j \log \frac{\hat{p}_j}{p_j} = -n \sum_{j=1}^{\alpha} (p_j + \delta_j) \left(\frac{\delta_j}{p_j} - \frac{1}{2} \frac{\delta_j^2}{p_j^2} \frac{1}{(1 + \xi_j)^2} \right) \\
&= -n \sum_{j=1}^{\alpha} \left(\frac{\delta_j^2}{p_j} + \frac{\delta_j^2}{p_j} \cdot \frac{\delta_j}{p_j} \cdot \left(-\frac{1}{(1 + \xi_j)^2} \right) \right) \quad \left(\sum_j \delta_j = 0 \right) \\
&= -\frac{n}{2} \sum_{j=1}^{\alpha} \frac{\delta_j^2}{p_j} + R_n,
\end{aligned}$$

where, by the central limit theorem, $\delta_j = O_P(n^{-1/2})$, and if α is fixed, then $\sum_{j=1}^{\alpha} |\delta_j|^3 = O_P(n^{-3/2})$. Hence,

$$R_n := n\sum_{j=1}^{\alpha} \hat{p}_j \frac{1}{(1+\xi_j)^2}\left(\frac{\delta_j}{p_j}\right)^2 \frac{\delta_j}{p_j} = O_P\left(n\sum_{j=1}^{\alpha}\frac{|\delta_j|^3}{p_j^2}\right) = O_P(n^{-1/2}) = o_P(1).$$

On the other hand,

$$n\sum_{j=1}^{\alpha}\frac{\delta_j^2}{p_j} = \sum_{j=1}^{\alpha}\frac{(k_j - np_j)^2}{np_j} \xrightarrow{d} \chi^2_{\alpha-1}$$

(Pearson's χ^2 statistic), so D(k, p) converges in distribution to $-\frac{1}{2}\chi^2_{\alpha-1}$, yielding (8.4).

Therefore, from (7.23) and (8.4),

$$\begin{aligned}
-\log Q_X(x^n) &= h(k_X, \hat{p}_X) + \frac{\alpha-1}{2}\log n + O(1) = h(k_X, p_X) \\
&\quad + \frac{\alpha-1}{2}\log n + O_P(1), \\
-\log Q_Y(y^n) &= h(k_Y, \hat{p}_Y) + \frac{\beta-1}{2}\log n + O(1) = h(k_Y, p_Y) \\
&\quad + \frac{\beta-1}{2}\log n + O_P(1), \\
-\log Q_{XY}(x^n, y^n) &= h(k_{XY}, \hat{p}_{XY}) + \frac{\alpha\beta-1}{2}\log n + O(1) = h(k_{XY}, p_{XY}) \\
&\quad + \frac{\alpha\beta-1}{2}\log n + O_P(1),
\end{aligned}$$

where k_X, k_Y, k_{XY} are the frequencies for X, Y, and (X, Y), respectively, and p_X, p_Y, p_{XY} are the corresponding probabilities. If X and Y are independent, then

$$K_n := h(k_X, p_X) + h(k_Y, p_Y) - h(k_{XY}, p_{XY})$$

equals 0 (Problem 84), and

$$\log Q_{XY}(x^n, y^n) - \log Q_X(x^n) - \log Q_Y(y^n) = -\frac{(\alpha-1)(\beta-1)}{2}\log n + O_P(1) \tag{8.5}$$

holds (Problem 84). Hence, as $n \to \infty$,

$$-\log Q_X(x^n) - \log Q_Y(y^n) < -\log Q_{XY}(x^n, y^n)$$

holds.

If X and Y are not independent, then, as $n \to \infty$, K_n/n converges in probability to the mutual information $I(X, Y)$, which is positive. Therefore, K_n grows faster than $\log n$, and

$$-\log Q_X(x^n) - \log Q_Y(y^n) > -\log Q_{XY}(x^n, y^n)$$

holds. ■

Proposition 28 guarantees that

$$J_n := \frac{1}{n} \log \frac{Q_{XY}(x^n, y^n)}{Q_X(x^n) Q_Y(y^n)} \tag{8.6}$$

is a valid measure of independence in the sense that non-positivity implies independence and positivity implies dependence.

Next, we consider independence testing and consistency in the continuous case.

Example 64 For a bivariate normal random variable (X, Y), we decide whether X and Y are independent from n paired observations. The function `Q.3` is the same as that created in Chap. 6.

```
1  # Data generation
2  n <- 100  # sample size
3  a <- rnorm(2) * 10                  # random coefficients
4  A <- a %*% t(a)                     # covariance matrix
5  z <- mvrnorm(n, c(0, 0), A)         # multivariate normal data
6  # Split into x, y
7  x <- z[, 1]
8  y <- z[, 2]
9  # Example calls
10 Q.3(x, 2, 2, c(0), matrix(1, 1, 1))                # using x
11 Q.3(y, 2, 2, c(0), matrix(1, 1, 1))                # using y
12 Q.3(z, 2, 2, c(0, 0), diag(2))                     # using (x,y)
```

We also experimented by replacing lines 3 to 5 above with the following:

```
z <- mvrnorm(n, c(0,0), diag(2))
```

Since `Q.3` is the negative log marginal likelihood, we decide independence if

```
Q.3(x, 2, 2, c(0), matrix(1, 1, 1)) +
Q.3(y, 2, 2, c(0), matrix(1, 1, 1)) -
Q.3(z, 2, 2, c(0, 0), diag(2))
```

is negative. The results (500 samples) are shown in Fig. 8.5c, d. In both cases, samples plotted above the red line were judged independent. Here we set $p = 0.5$ in (8.3). ■

We now note the following proposition.

Proposition 29 *Assume x^n and y^n are length-n sequences drawn independently from identical normal distributions (realizations of random variables X and Y), with priors given by (7.18). Then*

$$J_n = -\frac{1}{2}\log(1-\hat{\rho}_n^2) - \frac{1}{2n}\log n + O_P(1/n), \tag{8.7}$$

where

$$\hat{\rho}_n := \frac{\sum_{i=1}^n (x_i - \bar{x})(y_i - \bar{y})}{\sqrt{\sum_{i=1}^n (x_i - \bar{x})^2 \sum_{j=1}^n (y_j - \bar{y})^2}}$$

is the sample correlation coefficient.

Proof See the appendix at the end of the chapter.

Proposition 30 *Under the same assumptions as Proposition 29, the decision rule that declares X and Y independent if (8.7) is nonpositive, and dependent if it is positive, is consistent in the sense that the probability of a correct decision converges to 1 as n increases.*

Proof: If X and Y are independent, then $\rho = 0$, and with $X \sim N(\mu_X, \sigma_X^2)$, $Y \sim N(\mu_Y, \sigma_Y^2)$, we have

$$\hat{\rho}_n - \frac{1}{n}\sum_{i=1}^n \frac{x_i - \mu_X}{\sigma_X}\frac{y_i - \mu_Y}{\sigma_Y}$$

converging in probability to 0. By the central limit theorem,

$$\sqrt{n}\cdot\frac{1}{n}\sum_{i=1}^n \frac{x_i - \mu_X}{\sigma_X}\frac{y_i - \mu_Y}{\sigma_Y} \sim N(0,1) \quad (n \to \infty),$$

hence $n\hat{\rho}_n^2 \sim \chi_1^2$ as $n \to \infty$. Since in general $-\frac{1}{2}z \le -\frac{1}{2}\log(1-z) \le \frac{z}{2(1-z)}$, we have

$$-\frac{1}{2}\hat{\rho}_n^2 \le -\frac{1}{2}\log(1-\hat{\rho}_n^2) \le \frac{\hat{\rho}_n^2}{2(1-\hat{\rho}_n^2)},$$

and thus $-(1/2)\log(1-\hat{\rho}_n^2) = O_P(1/n)$. Therefore, in (8.7), the term $(1/2n)\log n$ dominates, and the expression diverges to $-\infty$.

If X and Y are not independent, then since we assumed normality, $\rho \neq 0$, and $\hat{\rho}_n^2$ converges in probability to $\rho^2 > 0$. Hence in (8.7), the term $-\frac{1}{2}\log(1-\hat{\rho}_n^2)$ dominates and the expression diverges to $+\infty$. ■

We omit the proof, but Propositions 28 and 30 also hold for conditional independence. That is,

$$Q_{XYZ}(x^n, y^n, z^n) + Q_Z(z^n) - Q_{XZ}(x^n, z^n) - Q_{YZ}(y^n, z^n)$$

being negative is equivalent to $X \perp\!\!\!\perp Y \mid Z$. The special case where the random variable Z is taken to be constant corresponds to Propositions 28 and 30.

From the above, we see that the 11 BNs in Fig. 3.6 can be distinguished using marginal likelihood. For example, using n samples from X, Y, Z, x^n, y^n, z^n, if

$$Q_{XYZ}(x^n, y^n, z^n) < \frac{Q_{XZ}(x^n, z^n) Q_{YZ}(y^n, z^n)}{Q_Z(z^n)},$$

we may regard this as $X \perp\!\!\!\perp Y \mid Z$, narrowing down to (a), (b), (c), (g). Further, if

$$Q_{XYZ}(x^n, y^n, z^n) < Q_{XZ}(x^n, z^n) Q_Y(y^n),$$

then it is one of (b) or (g), and so on. Including the other five cases, we obtain Table 8.1 (Problem 85). Thus, by examining all orderings, we effectively compare all 11 structures.

That is,

$$\begin{aligned}
&Q_X(x^n)Q_Y(y^n)Q_Z(z^n),\ Q_X(x^n)Q_Y(y^n, z^n),\ Q_Y(y^n)Q_{ZX}(z^n, x^n),\\
&Q_Z(z^n)Q_{XY}(x^n, y^n)\\
&\frac{Q_{ZX}(z^n, x^n)Q_{XY}(x^n, y^n)}{Q_X(x^n)},\ \frac{Q_{XY}(x^n, y^n)Q_{YZ}(y^n, z^n)}{Q_Y(y^n)},\ \frac{Q_{ZX}(z^n, x^n)Q_{XY}(x^n, y^n)}{Q_Z(z^n)}\\
&\frac{Q_Y(y^n)Q_Z(z^n)Q_{XYZ}(x^n, y^n, z^n)}{Q_{YZ}(y^n, z^n)},\ \frac{Q_Z(z^n)Q_X(x^n)Q_{XYZ}(x^n, y^n, z^n)}{Q_{ZX}(z^n, x^n)},\\
&\frac{Q_X(x^n)Q_Y(y^n)Q_{XYZ}(x^n, y^n, z^n)}{Q_{XY}(x^n, y^n)},\ Q_{XYZ}(x^n, y^n, z^n)
\end{aligned} \tag{8.8}$$

are each multiplied by prior structure probabilities $\pi_1, \ldots, \pi_{11} \geq 0$ ($\pi_1 + \cdots + \pi_{11} = 1$), and the structure with the largest resulting value is selected as the maximum a

Table 8.1 Structures to be compared (eight for each) when the order of X, Y, Z is fixed

Direction	(a)	(b)	(c)	(d)	(e)	(f)	(g)	(h)	(i)	(j)	(k)
$X \to Y \to Z$	✓	✓	✓	✓	✓	✓				✓	✓
$X \to Z \to Y$	✓	✓	✓	✓	✓		✓		✓		✓
$Y \to X \to Z$	✓	✓	✓	✓	✓	✓				✓	✓
$Y \to Z \to X$	✓	✓	✓	✓		✓	✓	✓			✓
$Z \to X \to Y$	✓	✓	✓	✓	✓		✓		✓		✓
$Z \to Y \to X$	✓	✓	✓	✓		✓	✓	✓			✓

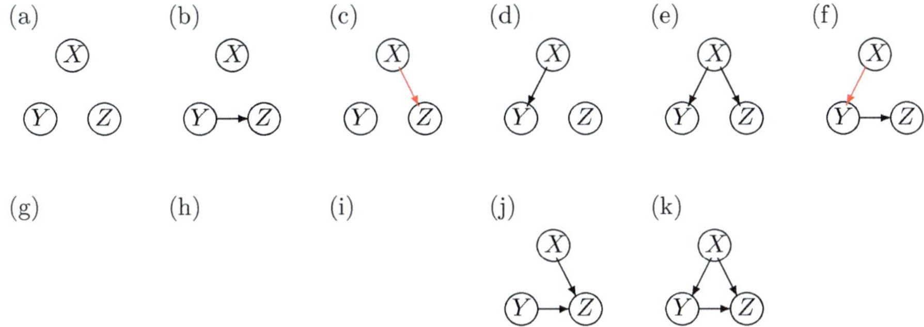

Fig. 8.6 Eight BNs assuming the order $X \to Y \to Z$. Compared with Fig. 3.6, the directions of the arrows in (**c**) and (**f**) are reversed, but they are Markov equivalent. There are no corresponding cases for (**g**), (**h**), or (**i**)

posteriori structure. A similar procedure can be applied to the Markov networks (MNs) in Fig. 3.7 (Problem 86).

When using information criteria, we first fixed an ordering (e.g., $X \to Y \to Z$) and then chose parent sets accordingly. Under such an ordering, the arrow directions of X, Z in Fig. 3.6c, and of X, Y in (f), are reversed (though they remain Markov equivalent and thus receive the same score). From Table 8.1, (g), (h), and (i) are excluded (Fig. 8.6). However, if we consider all six orderings, we will obtain a structure that is Markov equivalent to one of these 11 cases. Each can be expressed as a product of conditional probabilities under the order $X \to Y \to Z$. For example, (f) and (j) are, respectively,

$$Q_X(x^n) \cdot \frac{Q_{XY}(x^n, y^n)}{Q_X(x^n)} \cdot \frac{Q_{YZ}(y^n, z^n)}{Q_Y(y^n)}, \qquad Q_X(x^n) \cdot Q_Y(y^n) \cdot \frac{Q_{XYZ}(x^n, y^n, z^n)}{Q_{XY}(x^n, y^n)}.$$

8.4 When No Ordering Is Assumed

When no variable ordering is assumed, we can obtain the parent sets that maximize the marginal likelihood via the following procedure. Suppose we have n samples for each variable $X_1, \ldots, X_p$.

1. For each pair (X, S) with $X \notin S \subseteq \{X_1, \ldots, X_p\}$, compute $R(X \mid S) := \max_{U \subseteq S} Q(X \mid U)$. That is, fix X and start from $R(X \mid \{\}) = Q(X)$; then, while enlarging S, use the recursion

$$R(X \mid S) = \max\{\max_{Y \in S} R(X \mid S \backslash \{Y\}),\ Q(X \mid S)\}$$

to compute $R(X \mid \cdot)$ (Fig. 8.7a).

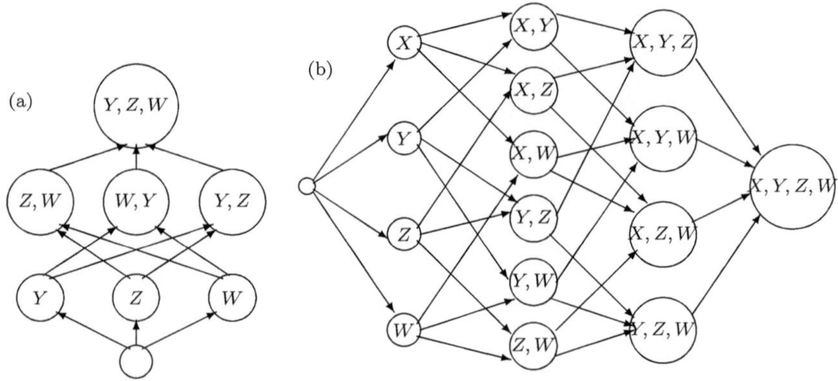

Fig. 8.7 (**a**) Procedure for computing $R(\cdot \mid \cdot)$ and (**b**) procedure for computing $T(X, Y, Z, W)$

2. Find the index ordering $X_1, \ldots, X_p$ that maximizes the product $\prod_{i=1}^{p} R(X_i \mid \{X_1, \ldots, X_{i-1}\})$. Starting from $T(\{\}) = 1$, enlarge S and use

$$T(S) = \max_{Y \in S} T(S \backslash \{Y\}) \cdot R(Y \mid S \backslash \{Y\}).$$

If we permute the indices to the optimal order, then

$$\begin{aligned} T(\{X_1, \ldots, X_p\}) &= T(\{X_1, \ldots, X_{p-1}\}) \cdot R(X_p \mid \{X_1, \ldots, X_{p-1}\}) = \cdots \\ &= \prod_{i=1}^{p} R(X_i \mid \{X_1, \ldots, X_{i-1}\}) \end{aligned}$$

holds (Fig. 8.7b).

The set U such that the final $R(X_i \mid \{X_1, \ldots, X_{i-1}\}) = Q(X_i \mid U)$ is achieved becomes the parent set of X_i [25, 26].

Example 65 For variables X, Y, Z, W, first compute $R(X \mid \cdot)$ (Fig. 8.7a). Let $R(X) = Q(X)$; then

$$\begin{cases} R(X \mid Y) = \max\{R(X), Q(X \mid Y)\} = \max\{Q(X), Q(X \mid Y)\}, \\ R(X \mid Z) = \max\{R(X), Q(X \mid Z)\} = \max\{Q(X), Q(X \mid Z)\}, \\ R(X \mid W) = \max\{R(X), Q(X \mid W)\} = \max\{Q(X), Q(X \mid W)\}. \end{cases}$$

$$\begin{cases} R(X \mid Z, W) = \max\{R(X \mid Z), R(X \mid W), Q(X \mid Z, W)\}, \\ R(X \mid W, Y) = \max\{R(X \mid W), R(X \mid Y), Q(X \mid W, Y)\}, \\ R(X \mid Y, Z) = \max\{R(X \mid Y), R(X \mid Z), Q(X \mid Y, Z)\}. \end{cases}$$

$$R(X \mid Y, Z, W)$$
$$= \max\{R(X \mid Z, W), R(X \mid W, Y), R(X \mid Y, Z), Q(X \mid Y, Z, W)\}.$$

Compute $R(Y \mid \cdot)$, $R(Z \mid \cdot)$, and $R(W \mid \cdot)$ similarly. Next,

$$T(X) = R(X), \quad T(Y) = R(Y), \quad T(Z) = R(Z), \quad T(W) = R(W)$$

$$\begin{cases} T(X, Y) &= \max\{T(X)R(Y \mid X),\ T(Y)R(X \mid Y)\} \\ &\vdots \\ T(Z, W) &= \max\{T(Z)R(W \mid Z),\ T(W)R(Z \mid W)\}. \end{cases}$$

$$\begin{cases} T(X, Y, Z) &= \max\{T(Y, Z)R(X \mid Y, Z),\ T(Z, X)R(Y \mid Z, X), \\ &\qquad T(X, Y)R(Z \mid X, Y)\} \\ &\vdots \\ T(Y, Z, W) &= \max\{T(Y, Z)R(W \mid Y, Z),\ T(Z, W)R(Y \mid Z, W), \\ &\qquad T(W, Y)R(Z \mid W, Y)\}. \end{cases}$$

$$T(X, Y, Z, W) = \max\{T(Y, Z, W)R(X \mid Y, Z, W), T(Z, W, X)R(Y \mid Z, W, X),$$
$$T(W, X, Y)R(Z \mid W, X, Y), T(X, Y, Z)R(W \mid X, Y, Z)\}$$

is obtained (Fig. 8.7b). Since the value $T(X, Y, Z, W)$ is a product of the $R(\cdot \mid \cdot)$ terms, the parent sets of each variable are determined. In practice, record the parent set when computing $T(\cdot)$; after the overall computation finishes, trace back those recorded parent sets.

Even if we do not maximize the marginal likelihood, we can run the same procedure by replacing it with minimizing AIC or BIC. For an implementation, see Problem 87.

However, even with methods that do not assume an ordering, Markov-equivalent structures receive the same score. If we need to determine edge directions, it is advisable to first infer an ordering among variables using LiNGAM as in Chap. 6, and then select parent sets based on the score.

8.5 BDeu

In general, the marginal likelihood is given by (7.16). In this section, suppose X and Y take values in $\{1, \ldots, \alpha\}$ and $\{1, \ldots, \beta\}$, respectively, and the frequencies of $(X, Y) = (i, j)$, $X = i$, $Y = j$ in $(x_1, y_1), \ldots, (x_n, y_n)$ are $n_{i,j}$, $n_{i,\cdot}$, and $n_{\cdot,j}$. Then, using some $\delta > 0$,

$$Q_X(x^n) = \left[\frac{\Gamma(n+\delta)}{\Gamma(\delta)}\right]^{-1} \prod_{i=1}^{\alpha} \frac{\Gamma(n_{i,\cdot}+\delta/\alpha)}{\Gamma(\delta/\alpha)},$$

$$Q_Y(y^n) = \left[\frac{\Gamma(n+\delta)}{\Gamma(\delta)}\right]^{-1} \prod_{j=1}^{\beta} \frac{\Gamma(n_{\cdot,j}+\delta/\beta)}{\Gamma(\delta/\beta)},$$

$$Q_{XY}(x^n, y^n) = \left[\frac{\Gamma(n+\delta)}{\Gamma(\delta)}\right]^{-1} \prod_{i=1}^{\alpha}\prod_{j=1}^{\beta} \frac{\Gamma(n_{i,j}+\delta/\alpha\beta)}{\Gamma(\delta/\alpha\beta)}$$

This is the Bayesian Dirichlet equivalent uniform (BDeu) method for setting marginal likelihoods [3]. It corresponds to choosing prior frequencies $a_{i,j} = \delta/(\alpha\beta)$, $a_{i,\cdot} = \delta/\alpha$, $a_{\cdot,j} = \delta/\beta$. For three or more variables, we can define scores analogously. The conditional marginal likelihood is

$$Q_{X|Y}(x^n, y^n) := \frac{Q_{XY}(x^n, y^n)}{Q_Y(y^n)} = \prod_{j=1}^{\beta} \left\{ \left[\frac{\Gamma(n_{\cdot,j}+\delta/\beta)}{\Gamma(\delta/\beta)}\right]^{-1} \prod_{i=1}^{\alpha} \frac{\Gamma(n_{i,j}+\delta/\alpha\beta)}{\Gamma(\delta/\alpha\beta)} \right\}. \tag{8.9}$$

In general,

$$\frac{\prod_{i=1}^{\alpha}\prod_{j=1}^{\beta}\Gamma(n_{i,j}+a_{i,j})}{\Gamma(\sum_{i=1}^{\alpha}\sum_{j=1}^{\beta}(n_{i,j}+a_{i,j}))} \cdot \frac{\Gamma(\sum_{i=1}^{\alpha}\sum_{j=1}^{\beta}(a_{i,j}))}{\prod_{i=1}^{\alpha}\prod_{j=1}^{\beta}\Gamma(a_{i,j})} \cdot \frac{\Gamma(\sum_{j=1}^{\beta}(n_{\cdot,j}+a_{\cdot,j}))}{\prod_{j=1}^{\beta}\Gamma(n_{\cdot,j}+a_{\cdot,j})}$$
$$\cdot \frac{\prod_{j=1}^{\beta}\Gamma(a_{\cdot,j})}{\Gamma(\sum_{j=1}^{\beta}(a_{\cdot,j}))}$$

holds for $Q_{X|Y}(x^n, y^n)$. The fact that $a_{i,j}, a_{i,\cdot}, a_{\cdot,j}$ are constant with respect to i, j is shared by BDeu and the prior frequencies considered so far. However, under Jeffreys' prior,

$$a_{i,j} = a_{i,\cdot} = a_{\cdot,j} = a > 0,$$

whereas BDeu sets $a_{i,j} = \dfrac{\delta}{\alpha\beta}$, so that

$$a_{\cdot,j} = \sum_{i=1}^{\alpha} a_{i,j} = \frac{\delta}{\beta}, \quad a_{i,\cdot} = \sum_{j=1}^{\beta} a_{i,j} = \frac{\delta}{\alpha}.$$

Writing out the case of Jeffreys' prior explicitly with $a = 1/2$ gives

$$Q_{X|Y}(x^n \mid y^n) = \prod_{j=1}^{\beta} \left\{ \frac{\Gamma(\alpha/2)}{\Gamma(n_{i,j} + \alpha/2)} \prod_{i=1}^{\alpha} \frac{\Gamma(n_{i,j} + 1/2)}{\Gamma(1/2)} \right\} \tag{8.10}$$

(Problem 89).

Because BDeu is adopted in the CRAN package `bnlearn` [22], it is more widely used than AIC, BIC, or Jeffreys' method. However, as we discuss below, it suffers from a critical issue.

Let S be a parent set for X, and let $H(X \mid S)$ and $d(X, S)$ denote the empirical entropy and the number of parameters, respectively. Using these, define AIC and BIC as

$$AIC(X, S) = H(X \mid S) + d(X, S), \quad BIC(X, S) = H(X \mid S) + \frac{d(X, S)}{2} \log n.$$

Then

$$AIC(X, S) > AIC(X, S'), \quad S \subset S' \Longrightarrow H(X \mid S) > H(X \mid S') \tag{8.11}$$

$$BIC(X, S) > BIC(X, S'), \quad S \subset S' \Longrightarrow H(X \mid S) > H(X \mid S') \tag{8.12}$$

hold (Problem 90). We refer to this as the *regularity* of information criteria.

If regularity fails–i.e., if $H(X \mid S') > H(X \mid S)$ (so S predicts X better than S'), yet the score (AIC or BIC) is smaller for S'–then the better predictive model is not selected, and a more complex, redundant model is mistakenly chosen. Such inversions can cause serious problems:

- The model becomes unnecessarily complex, leading to *overfitting*. There is a risk of inferring spurious causal relationships.
- In the model selection process, adding parents does not necessarily correspond to improving the likelihood (entropy), making it difficult to design search algorithms (e.g., pruning becomes impossible) [38].

Therefore, to ensure consistency and interpretability of information criteria in structure learning, it is desirable that such regularity holds.

BDeu does not satisfy regularity [33].

Example 66 Let X and Y take α and β distinct values, respectively, and consider $n \geq 2$ with $x^n = \underbrace{1, \ldots, 1}_{n}$, $y^n = \underbrace{1, \ldots, 1}_{n}$. We compute the marginal likelihoods. Since the occurrence of X does not depend on Y, it should be that $Q_X(x^n) \geq$

$Q_{X|Y}(x^n \mid y^n)$. Using Jeffreys' prior,

$$Q_X(x^n) = \frac{\Gamma(\alpha/2)}{\Gamma(n+\alpha/2)} \cdot \frac{\Gamma(n+1/2)}{\Gamma(1/2)},$$

$$Q_Y(y^n) = \frac{\Gamma(\beta/2)}{\Gamma(n+\beta/2)} \cdot \frac{\Gamma(n+1/2)}{\Gamma(1/2)},$$

$$Q_{XY}(x^n, y^n) = \frac{\Gamma(\alpha\beta/2)}{\Gamma(n+\alpha\beta/2)} \cdot \frac{\Gamma(n+1/2)}{\Gamma(1/2)}, \tag{8.13}$$

$$Q_{X|Y}(x^n \mid y^n) = \left[\frac{\Gamma(\alpha\beta/2)}{\Gamma(n+\alpha\beta/2)}\right]\left[\frac{\Gamma(\beta/2)}{\Gamma(n+\beta/2)}\right]^{-1} \le Q_X(x^n) \tag{8.14}$$

are obtained. Here, (8.13) treats (X, Y) as a single variable taking $\alpha\beta$ values. The inequality in (8.14) follows because

$$a_n := \frac{\Gamma(n+\alpha\beta/2)\Gamma(n+1/2)}{\Gamma(n+\alpha/2)\Gamma(n+\beta/2)} \cdot \frac{\Gamma(\alpha/2)\Gamma(\beta/2)}{\Gamma(\alpha\beta/2)\Gamma(1/2)}, \qquad n \ge 0$$

is monotone nondecreasing with $a_0 = 1$ (Problem 91). Moreover, the equality $a_{n+1} \ge a_n$ does not hold when $n \ne 0$ (Problem 91). On the other hand, under BDeu,

$$Q_X(x^n) = \frac{\Gamma(\delta)}{\Gamma(n+\delta)} \cdot \frac{\Gamma(n+\delta/\alpha)}{\Gamma(\delta/\alpha)}, \tag{8.15}$$

$$Q_Y(y^n) = \frac{\Gamma(\delta)}{\Gamma(n+\delta)} \cdot \frac{\Gamma(n+\delta/\beta)}{\Gamma(\delta/\beta)}, \tag{8.16}$$

$$Q_{XY}(x^n, y^n) = \frac{\Gamma(\delta)}{\Gamma(n+\delta)} \cdot \frac{\Gamma(n+\delta/(\alpha\beta))}{\Gamma(\delta/(\alpha\beta))},$$

$$Q_{X|Y}(x^n \mid y^n) = \left[\frac{\Gamma(n+\delta/(\alpha\beta))}{\Gamma(\delta/(\alpha\beta))}\right]\left[\frac{\Gamma(n+\delta/\beta)}{\Gamma(\delta/\beta)}\right]^{-1} \ge Q_X(x^n) \tag{8.17}$$

holds. The inequality in (8.17) follows because

$$b_n := \frac{\Gamma(n+\delta)}{\Gamma(n+\delta/\alpha)} \frac{\Gamma(n+\delta/(\alpha\beta))}{\Gamma(n+\delta/\beta)} \cdot \frac{\Gamma(\delta/\alpha)\Gamma(\delta/\beta)}{\Gamma(\delta)\Gamma(\delta/(\alpha\beta))}, \qquad n \ge 0$$

is monotone nondecreasing with $b_0 = 1$ (Problem 91). Moreover, the equality $b_{n+1} \ge b_n$ does not hold when $n \ne 0$ (Problem 91). ■

Example 67 In the datasets shown in (a) and (b) below, we have $H(X \mid Y) = H(X \mid Y, Z) = 0$. In other words, X is completely determined by Y, and adding Z does not decrease the conditional entropy any further.

From the perspective of the regularity of information criteria, the parent set should therefore be $\{Y\}$, while the empty set or the excessive sets $\{Z\}$ and $\{Y, Z\}$ should be avoided.

When evaluated with the marginal likelihood based on Jeffreys' prior, the following holds:

$$Q_{X|Y}(x^n \mid y^n) > Q_{X|YZ}(x^n \mid y^n, z^n),$$

i.e., adding an unnecessary parent variable Z lowers (worsens) the score, so Jeffreys' method preserves regularity. In contrast, when evaluated by the BDeu score, extending (8.9) to $Q_{X|YZ}(x^n, y^n, z^n)$ gives

$$\prod_{j=1}^{\beta}\prod_{k=1}^{\gamma}\left\{\left[\frac{\Gamma(n_{\cdot,j,k}+\delta/\beta\gamma)}{\Gamma(\delta/(\beta\gamma))}\right]^{-1}\prod_{i=1}^{\alpha}\frac{\Gamma(n_{i,j,k}+\delta/\alpha\beta\gamma)}{\Gamma(\delta/\alpha\beta\gamma)}\right\},$$

and we have

$$Q_{X|Y}(x^n \mid y^n) < Q_{X|YZ}(x^n \mid y^n, z^n)$$

(Problem 93), which means that $\{Y, Z\}$, which should not be chosen, attains a higher score than $\{Y\}$. In other words, *regularity is violated.* This behavior of BDeu indicates a tendency to include redundant variables as parents even when the data satisfy $H(X \mid Y) = 0$. A concrete example is shown below.

(a)

X	Y	Z
1	1	2
2	2	1
1	1	2
2	2	1

(b)

X	Y	Z
1	1	1
2	2	1
1	3	2
2	4	2
1	1	1
2	2	1
1	3	2
2	4	2

Here, (a) corresponds to $(\alpha, \beta, \gamma) = (2, 2, 2)$ and (b) to $(\alpha, \beta, \gamma) = (2, 4, 2)$. In case (a), using $\Gamma(x+1) = x\Gamma(x)$ for $x > 0$, we have

$$\frac{\Gamma(4/2)}{\Gamma(4+4/2)}\left[\frac{\Gamma(2/2)}{\Gamma(4+2/2)}\right]^{-1} = \frac{1}{5} > \frac{1}{7} = \frac{\Gamma(8/2)}{\Gamma(4+8/2)}\left[\frac{\Gamma(4/2)}{\Gamma(4+4/2)}\right]^{-1} \tag{8.18}$$

$$\left[\frac{\Gamma(4+1/2)}{\Gamma(1/2)}\right]^{-1}\frac{\Gamma(4+1/4)}{\Gamma(1/4)}=\frac{3\cdot 13}{2^4\cdot 7}<\frac{5\cdot 17}{2^4\cdot 13}=\left[\frac{\Gamma(4+1/4)}{\Gamma(1/4)}\right]^{-1}\frac{\Gamma(4+1/8)}{\Gamma(1/8)} \tag{8.19}$$

so, under Jeffreys' prior ($a = 1/2$) and under BDeu, the respective inequalities between $Q_{X|Y}(x^n, y^n)$ and $Q_{X|YZ}(x^n, y^n, z^n)$ are $>$ and $<$. Case (b) can be verified similarly (Problem 93). In the BDeu calculation, as in (b), increasing the number of states tends to make such "extraneous variables" more advantageous. ■

8.6 Learning Forest Structures

Thus far, we have addressed the problem of constructing a BN from samples. For p variables, the computation time grows exponentially in p. In fact, even if the ordering is known, exploring $\pi_k \subseteq \{1, \ldots, k-1\}$ requires comparing 2^{k-1} candidate parent sets, so we must compare $\sum_{k=1}^{p}(2^{k-1}) = 2^p - 1$ structures in total. If we also need to search for an ordering using a score, the computational cost becomes even larger.

In the Chow–Liu algorithm from Chap. 2, if the true mutual information $I(X_i, X_j)$ were available, we would obtain an approximation to the distribution. When we have only data, and X, Y take values in $\{1, \ldots, \alpha\}$ and $\{1, \ldots, \beta\}$, respectively, with frequencies in $(x_1, y_1), \ldots, (x_n, y_n)$ denoted by $n_{i,\cdot}$ for $X = i$, $n_{\cdot,j}$ for $Y = j$, and $n_{i,j}$ for $(X, Y) = (i, j)$, it is common to estimate the mutual information by

$$I_n := \sum_i \sum_j \frac{n_{i,j}}{n} \log\left(\frac{n_{i,j}}{n} \Big/ \frac{n_{i,\cdot}}{n}\frac{n_{\cdot,j}}{n}\right). \tag{8.20}$$

In R, one can implement I_n as follows:

```
I.n <- function(x, y) {
    z <- table(x, y)  # joint frequency table of X, Y
    size <- dim(z)    # table size
    n <- length(x)    # sample size
    S <- 0            # initialize mutual information
    for (i in 1:size[1]) {
        u <- sum(z[i,])        # frequency of X = i
        for (j in 1:size[2]) {
            v <- sum(z[, j])  # frequency of Y = j
            if (z[i, j] != 0) {
                S <- S + z[i, j] / n * log( z[i, j] * n / (u * v) )
                # compute mutual information
            }
        }
    }
    return(S)
}
```

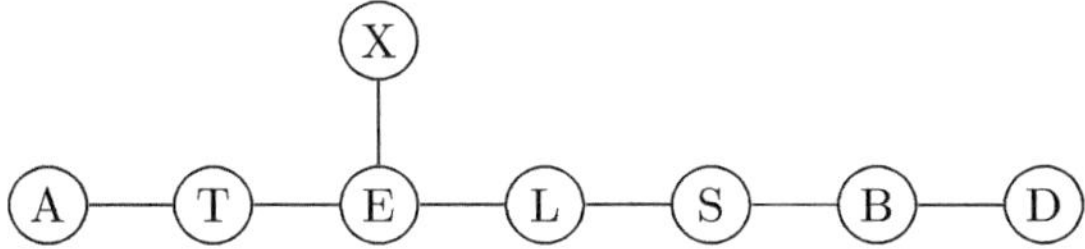

Fig. 8.8 Maximum spanning tree obtained by estimating mutual information for the Asia dataset using the function `I.n` (maximum likelihood method) and applying Kruskal's algorithm

Example 68 `Asia` is a dataset with $p = 8$ variables and $n = 5000$ samples, as shown in Table 1.1.

We estimated mutual information from the data using `I.n`, formed a weight matrix, and ran `kruskal`. The resulting `edgelist` was visualized with `igraph` (Fig. 8.8).

```
library(igraph)
library(bnlearn)
# Load data
df <- asia
n <- nrow(df)  # number of samples
p <- ncol(df)  # number of variables
# Convert to matrix form
x <- matrix(nrow = n, ncol = p)
for (i in 1:p) {
    x[, i] <- df[[i]]
}
# Build the mutual information matrix
w <- matrix(0, nrow = p, ncol = p)  # initialize
for (i in 1:(p - 1)) {
    for (j in (i + 1):p) {
        w[i, j] <- I.n(x[, i], x[, j])  # compute mutual information
        w[j, i] <- w[i, j]
    }
}
# Construct the Chow -- Liu tree (maximum-weight spanning tree)
edge.list <- kruskal(w)  # apply Kruskal's algorithm
print(edge.list)         # show the edge list
# Plot the graph
g <- graph_from_edgelist(edge.list, directed = FALSE)  # use igraph
V(g)$label <- colnames(df)  # label nodes
plot(g, main = "Chow -- Liu Algorithm (I.n)")  # draw the graph
```

■

In this section, we extend the Chow–Liu algorithm to the problem of constructing a forest [30]. We still apply Kruskal's algorithm, but the focus is on improving how mutual information is estimated.

Up to now, we have assumed X, Y take values in $\{1, \ldots, \alpha\}$ and $\{1, \ldots, \beta\}$ and constructed an estimator of $I(X, Y)$ from the samples $(x_1, y_1), \ldots, (x_n, y_n)$ as in (8.20). A merit of this approach is that I_n converges in probability to $I(X, Y)$ as $n \to \infty$.

However, even when $X \perp\!\!\!\perp Y$, for finite n, the value of I_n remains positive (though it decreases toward 0 as n grows). Hence, it cannot be used directly for independence testing. Here, we adopt a different estimator of mutual information, redefining it by modifying (8.6):

$$J_n := \max\left\{\frac{1}{n}\log\frac{Q_{XY}(x^n, y^n)}{Q_X(x^n)Q_Y(y^n)}, 0\right\}. \tag{8.21}$$

An R implementation is as follows:

```
J.n <- function(x,y){
  x <- as.numeric(x)
  alpha <- length(table(x))
  y <- as.numeric(y)
  beta <- length(table(y))
  z <- (y-1) * alpha + x
  value <- Q.4(x, alpha) + Q.4(y, beta) - Q.4(z, alpha*beta)
  return(max(value,0))
}
```

Here the marginal likelihood is computed by the following function `Q.4`:

```
1  Q.4 <- function(x, alpha){
2    x <- as.numeric(x)
3    n <- length(x)
4    nc <- 0
5    cc <- rep(0, alpha)
6    log.q <- 0
7    for(i in 1:n){
8      log.q <- log.q - log ( (cc[x[i]]+0.5) / (nc + alpha/2))
9      cc[x[i]] <- cc[x[i]] + 1
10     nc <- nc +1
11   }
12   return(log.q / n)
13 }
```

For small n, one can construct this using `Q.1` from Chap. 6, but since factorials are involved, it is necessary to work on the log scale and accumulate as in line 8 above.

By Proposition 28, we have

$$J_n = 0 \Longleftrightarrow X \perp\!\!\!\perp Y$$

with probability 1. Between the estimators I_n and J_n, there is the relation

$$J_n = \max\{I_n - \frac{1}{2n}(\alpha - 1)(\beta - 1)\log n + O(1/n), 0\} \tag{8.22}$$

(Problem 96). Here, the second term $(\alpha - 1)(\beta - 1)\log n/(2n)$ penalizes model complexity (larger numbers of states), and it can be interpreted as controlling overfitting to the training data.

The forest construction algorithm published by the author in 1993 [30] also applied, not (8.21) but [32], the value in (8.22) as the mutual information in the Chow–Liu algorithm. In R, for example:

```
J.n <- function(x,y){
  n <- length (x)
  size <- dim(table(x,y))
  value <- I.n(x,y) - (size[1]-1) * (size[2]-1) / 2 / n * log(n)
  return(max(value,0))
}
```

The resulting BN is a forest, not a tree, because independent vertex sets are not connected by paths. Moreover, when using J_n, the order in which edges are joined differs from simply truncating the tree obtained with I_n. The second term in (8.22) serves to correct for overfitting. When $(\alpha-1)(\beta-1)$ is large, the tendency to overfit is stronger.

In the program of Example 68, we used `I.n` to obtain Fig. 8.8; if we replace it by `J.n` in (8.21) and run again, we obtain Fig. 8.9. The node `Asia` is separated from the other vertices. This means that, in the data, `Asia` has low mutual information with the other variables and is regarded as independent in model selection. Using `J.n` from (8.22) yields the same result. However, since all variables are binary here, $(\alpha-1)(\beta-1)$ takes the same value, and thus the order in which edges are joined is the same (Fig. 8.10).

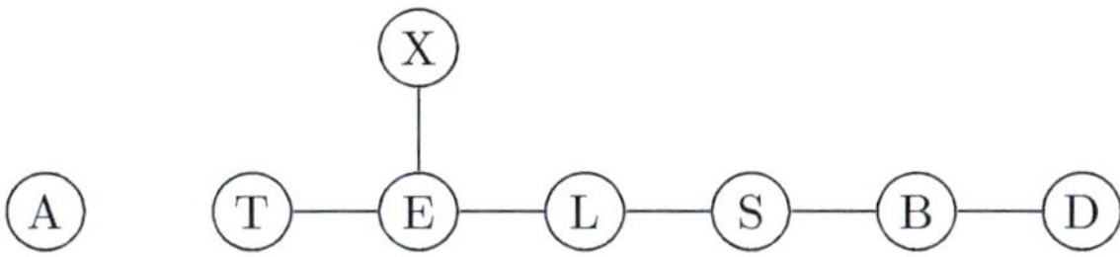

Fig. 8.9 Maximum spanning tree obtained by estimating mutual information with function `J.n` (Bayesian method) for the Asia dataset and applying Kruskal's algorithm. Unlike Fig. 8.8, Asia is separated

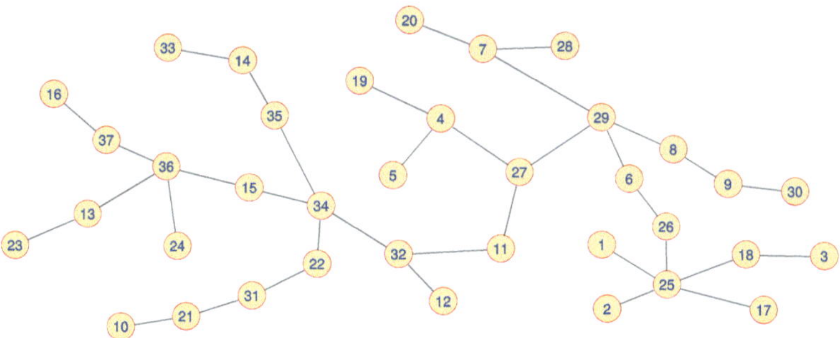

Fig. 8.10 Applying the Chow–Liu algorithm to the Alarm dataset with $p = 37$. In this example, the forests obtained using I_n and using J_n from (8.21) coincide

By Propositions 29 and 30, even under the Gaussian assumption, we can define

$$I_n := -\frac{1}{2}\log(1-\hat{\rho}_n^2), \qquad J_n := -\frac{1}{2}\log(1-\hat{\rho}_n^2) - \frac{1}{2n}\log n.$$

In R, these can be written as follows:

```
I.n <- function(x,y) {
  rho <- cov(x, y) / sqrt(var(x) * var(y))
  return(-0.5 * log(1-rho^2))
}
```

```
J.n <- function(x,y){
  n <- length(x)
  value <- I.n(x,y) - 0.5 * log (n) / n
  return (max(value,0))
}
```

From the foregoing discussion, as in the discrete case, J_n converges to the true mutual information $I(X, Y)$ as $n \to \infty$, and $J_n = 0$ if and only if X and Y are independent (Problem 98).

Appendix

The proof of Proposition 27 below follows Theorem 2 of [40].

Proof of Proposition 27

For any disjoint $S, T, U \subseteq \{1, \ldots, p\}$ ($S, T \neq \emptyset$), we say G is a BN if

$$S \perp\!\!\!\perp_G T \mid U \Longrightarrow X_S \perp\!\!\!\perp X_T \mid X_U \tag{8.23}$$

holds, and if deleting any edge from E makes this implication fail (i.e., G is minimal). (See Chap. 2.)

We prove by induction that for each $2 \leq k \leq p$, with $V = \{1, \ldots, k\}$ and $E = \cup_{j=1}^{k}\{(i, j) \mid i \in \pi_j\}$, the graph is a BN for $X_1, \ldots, X_k$. For $k = 2$, the disjoint $S, T, U \subseteq \{1, 2\}$ ($S, T \neq \emptyset$) must have $U = \emptyset$ and either $S = \{1\}, T = \{2\}$ or $S = \{2\}, T = \{1\}$, so (8.23) holds. Also, since $E = \emptyset$, no edge can be removed, so minimality holds.

Assume $V = \{1, \ldots, k-1\}$ and $E = \cup_{j=1}^{k-1}\{(i, j) \mid i \in \pi_j\}$ form a BN for $X_1, \ldots, X_{k-1}$. When we add vertex k and edges (j, k) for $j \in \pi_k$ to V and E, we will show

$$S \perp\!\!\!\perp_G T \mid U \Longrightarrow X_S \perp\!\!\!\perp X_T \mid X_U, \tag{8.24}$$

$$(S \cup k) \perp\!\!\!\perp_G T \mid U \Longrightarrow X_{S\cup k} \perp\!\!\!\perp X_T \mid X_U, \tag{8.25}$$

$$S \perp\!\!\!\perp_G T \mid (U \cup k) \Longrightarrow X_S \perp\!\!\!\perp X_T \mid X_{U\cup k}. \tag{8.26}$$

This implies that (8.24) holds for $V = \{1, \ldots, k\}$ and $E = \cup_{j=1}^{k}\{(i, j) \mid i \in \pi_j\}$.

For (8.24), we only add edges pointing into k, which create collisions at k, so paths between S and T are unaffected. Adding X_k does not affect conditional independences among $X_1, \ldots, X_{k-1}$; hence (8.24) holds.

For (8.25), observe by assumption that

$$X_k \perp\!\!\!\perp X_{\overline{\pi}_k} \mid X_{\pi_k}. \tag{8.27}$$

Here, $\overline{A}$ denotes the complement of A in $\{1, \ldots, k-1\}$. Let

$$\pi_* := \pi_k \cap \overline{S \cup T \cup U}, \qquad \pi^* := \overline{\pi}_k \cap \overline{S \cup T \cup U} = \overline{\pi_k \cup S \cup T \cup U}$$

(see Fig. 8.11a). Applying (3.6) with

$$A := k, \qquad B := (T \cup \pi^*) \cap \overline{\pi}_k, \qquad C := \pi_k, \qquad D := \overline{T \cup \pi^*} \cap \overline{\pi}_k,$$

(8.27) gives

$$X_k \perp\!\!\!\perp X_{(T\cup\pi^*)\cap\overline{\pi}_k} \mid X_{\pi_k \cup \overline{T\cup\pi^*}\cap\overline{\pi}_k}.$$

Note that $T \cap \pi_k = \emptyset$. Indeed, if $j \in \pi_k \cap T$, there is a directed edge of length 1 between j and k, so unless U contains both j and k, this contradicts $(S \cup k) \perp\!\!\!\perp_G T \mid U$ on the left-hand side of (8.25). Therefore,

$$\begin{aligned}
(T \cup \pi^*) \cap \overline{\pi}_k &= T \cup (\pi^* \cap \pi_k) = T \cup \pi^*, \\
\overline{T \cup \pi^*} &= \overline{T} \cap \overline{\pi^*} = \overline{T} \cap (\pi_k \cup S \cup T \cup U) = S \cup U \cup \pi_k, \\
\pi_k \cup \overline{T \cup \pi^*} \cap \overline{\pi}_k &= \pi_k \cup \overline{T \cup \pi^*} = S \cup U \cup \pi_k,
\end{aligned}$$

and hence

$$X_k \perp\!\!\!\perp X_{T\cup\pi^*} \mid X_{S\cup U\cup\pi_k}$$

(see Fig. 8.11a). Applying (3.5) further yields

$$X_k \perp\!\!\!\perp X_T \mid X_{S\cup U\cup\pi_k}. \tag{8.28}$$

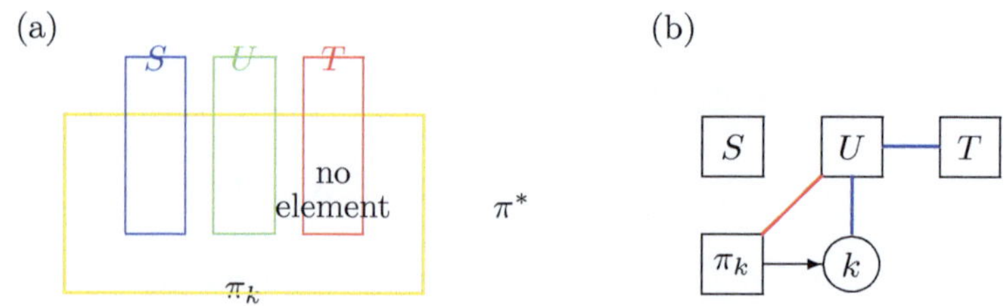

Fig. 8.11 Proof of Proposition 27. (**a**) Since $T \cap \pi_k = \{\}$, it follows that $\overline{T \cup \pi^*} = S \cup U \cup \pi_k$. (**b**) If $k \perp\!\!\!\perp_G T \mid U$, then it must also hold that $\pi_k \perp\!\!\!\perp_G T \mid U$; otherwise, there exists a path from k to T through π_k

On the other hand, from the assumption $k \perp\!\!\!\perp_G T \mid U$, we need $\pi_k \perp\!\!\!\perp_G T \mid U$; otherwise, since all edges point from π_k to k, there would exist a vertex in T not d-separated from k by U (Fig. 8.11b). Together with $S \perp\!\!\!\perp_G T \mid U$, we have $(S \cup \pi_k) \perp\!\!\!\perp_G T \mid U$, and by (8.24),

$$X_{S \cup \pi_k} \perp\!\!\!\perp X_T \mid X_U. \tag{8.29}$$

Applying (3.7) with

$$A := T, \qquad B := S \cup \pi_k, \qquad C := U, \qquad D := k$$

to (8.28) and (8.29) gives

$$X_{S \cup k \cup \pi_k} \perp\!\!\!\perp X_T \mid X_U.$$

Applying (3.5) yields the right-hand side of (8.25).

For (8.26), we assume $S \perp\!\!\!\perp_G T \mid (U \cup k)$. Then, any path between $s \in S$ and $t \in T$ has a collision at k, so k does not act as a d-separator. Therefore, either $k \perp\!\!\!\perp_G S \mid U$ or $k \perp\!\!\!\perp_G T \mid U$ must hold. Without loss of generality, assume $k \perp\!\!\!\perp_G T \mid U$. Then, by (3.8),

$$S \perp\!\!\!\perp_G T \mid U \quad \text{and} \quad k \perp\!\!\!\perp_G T \mid U \iff (S \cup k) \perp\!\!\!\perp_G T \mid U.$$

Applying (8.25) then yields $X_{S \cup k} \perp\!\!\!\perp X_T \mid X_U$. Finally, applying (3.6) gives $X_S \perp\!\!\!\perp X_T \mid X_{U \cup k}$, establishing (8.26).

Thus, (8.24) holds for any disjoint $S, T, U \subseteq \{1, \ldots, k\}$.

Moreover, by construction, the graph is a boundary DAG (Sect. 8.1). Deleting any edge breaks (8.24), so minimality also holds, completing the proof. ■

Proof of Proposition 29

$$Q_X = \pi^{-n/2}\frac{\Gamma(\nu_n/2)}{\Gamma(\nu_0/2)}\frac{\lambda_{X,0}^{\nu_0/2}}{\lambda_{X,n}^{\nu_n/2}}\left(\frac{\kappa_0}{\kappa_n}\right)^{1/2}, \qquad Q_Y = \pi^{-n/2}\frac{\Gamma(\nu_n/2)}{\Gamma(\nu_0/2)}\frac{\lambda_{Y,0}^{\nu_0/2}}{\lambda_{Y,n}^{\nu_n/2}}\left(\frac{\kappa_0}{\kappa_n}\right)^{1/2}$$

$$Q_{XY} = \pi^{-n}\frac{\Gamma_2(\nu_n/2)}{\Gamma_2(\nu_0/2)}\frac{\det(\Lambda_0)^{\nu_0/2}}{\det(\Lambda_n)^{\nu_n/2}}\left(\frac{\kappa_0}{\kappa_n}\right)^{1}$$

$$\lambda_{X,n} = \lambda_{X,0}+(\bar{x}-\mu_{X,0})^2+\sum_{i=1}^{n}(x_i-\bar{x})^2, \qquad \lambda_{Y,n} = \lambda_{Y,0}+(\bar{y}-\mu_{Y,0})^2+\sum_{i=1}^{n}(y_i-\bar{y})^2$$

$$\lambda_n = \lambda_0 + (\bar{x}-\mu_{X,0})(\bar{y}-\mu_{Y,0}) + \sum_{i=1}^{n}(x_i-\bar{x})(y_i-\bar{y})$$

$$\Lambda_0 = \begin{bmatrix} \lambda_{X,0} & \lambda_0 \\ \lambda_0 & \lambda_{Y,0} \end{bmatrix}, \qquad \Lambda_n = \begin{bmatrix} \lambda_{X,n} & \lambda_n \\ \lambda_n & \lambda_{Y,n} \end{bmatrix}$$

Let

$$\hat{\rho}_n^2 - \frac{\lambda_n^2}{\lambda_{X,n}\lambda_{Y,n}}$$

converge in probability to 0. It suffices then to show

$$-\log\frac{\Gamma_2(\nu_n/2)}{\Gamma_2(\nu_0/2)} = \frac{1}{2}\log n + O(1). \tag{8.30}$$

First note

$$\frac{\Gamma_2(\frac{\nu_n}{2})}{\Gamma(\frac{\nu_n}{2})} = \pi^{1/2}\frac{\Gamma(\frac{\nu_n}{2})\Gamma(\frac{\nu_n-1}{2})}{\Gamma(\frac{\nu_n}{2})^2} = \pi^{1/2}\frac{\Gamma(\frac{\nu_n-1}{2})}{\Gamma(\frac{\nu_n}{2})}.$$

Since $\log\Gamma(x) = \log(\int_0^\infty t^{x-1}e^{-t}dt)$ is twice differentiable and convex, we have Gautschi's inequality $\Gamma(\frac{1}{2}u + \frac{1}{2}v) < \Gamma(u)^{1/2}\Gamma(v)^{1/2}$. Substituting $(u, v) = (\frac{\nu_n-2}{2}, \frac{\nu_n}{2})$ and $(\frac{\nu_n-1}{2}, \frac{\nu_n+1}{2})$ yields

$$\Gamma(\tfrac{\nu_n-1}{2}) < \Gamma(\tfrac{\nu_n-2}{2})^{1/2}\Gamma(\tfrac{\nu_n}{2})^{1/2} = \left(\tfrac{\nu_n-2}{2}\right)^{-1/2}\Gamma(\tfrac{\nu_n}{2}),$$

$$\Gamma(\tfrac{\nu_n}{2}) < \Gamma(\tfrac{\nu_n-1}{2})^{1/2}\Gamma(\tfrac{\nu_n+1}{2})^{1/2} = \left(\tfrac{\nu_n-1}{2}\right)^{1/2}\Gamma(\tfrac{\nu_n-1}{2}).$$

Hence,

$$\sqrt{\frac{\nu_n - 2}{2}} < \frac{\Gamma(\frac{\nu_n}{2})}{\Gamma(\frac{\nu_n - 1}{2})} < \sqrt{\frac{\nu_n - 1}{2}},$$

and therefore,

$$\frac{1}{2}\log(\nu_n - 2) - \frac{1}{2}\log 2\pi < \log \frac{\Gamma(\frac{\nu_n}{2})^2}{\Gamma_2(\frac{\nu_n}{2})} < \frac{1}{2}\log(\nu_n - 1) - \frac{1}{2}\log 2\pi .$$

This establishes (8.30). ■

Problems 79–100

79. Analogously to the hill-climbing (BIC) structure learning on the `alarm` dataset in Sect. 7.1, obtain `edges` for the `asia` dataset and visualize the learned BN with `igraph`.

```
library(igraph)
# build an igraph object
g <- graph_from_edgelist(edges, directed = TRUE)
# plot with vertex labels
plot(g, vertex.label = c("A", "S", "T", "L", "B", "E", "X", "D"),
     label = V(g)$name)
```

Then replace `BIC` with `AIC` and rerun. Compare the structures and describe the differences (e.g., AIC typically prefers denser graphs than BIC).

80. In Example 62, run `table(X.1, X.2, X.3)`, `ftable(X.1, X.2, X.3)`, and `ftable(X.2, X.3, X.1)`. Focusing on the layout and readability of the printed tables, explain the behavioral differences between `table` and `ftable` (e.g., flat contingency layout and variable order effects).
81. For two candidate parent sets with empirical entropy and parameter counts (H, d) and (H', d'), show that

$$H + \frac{d}{2}\log n \le H' + \frac{d'}{2}\log n, \qquad H + d \ge H' + d'$$

can simultaneously hold only if $d \le d'$. Conclude that, because BIC penalizes complexity more than AIC, BIC tends to select simpler structures.

82. The function `IC.min` in Sect. 7.2 is implemented for continuous variables. Rewrite it for discrete variables, apply it to the `asia` dataset, and compare the learned structure with that obtained by hill climbing in Sect. 7.1.

83. The function Q below takes an $n \times p$ matrix and returns the negative (per-sample) log marginal likelihood for the p variables. In the example, it evaluates three of the eleven Markov-equivalent BN scores for three variables. Extend it to evaluate all remaining eight and identify which BN maximizes the marginal likelihood.

```
Q <- function(x) {
  x <- as.matrix(x)
  n <- nrow(x)
  p <- ncol(x)
  m <- array(p)
  M <- 1
  for (j in 1:p) {
    m[j] <- length(table(x[, j]))
    M <- M * m[j]
  }
  cc <- array(0, dim = M)
  q <- 0
  for (i in 1:n) {
    z <- x[i, 1] - 1
    if (p > 1) {
      for (j in 2:p) {
        z <- z * m[j] + (x[i, j] - 1)
      }
    }
    z <- z + 1
    q <- q - log((cc[z] + 0.5) / (i - 1 + 0.5 * M))
    cc[z] <- cc[z] + 1
  }
  return(q / n)
}
n <- 1000
u <- rbinom(n, 1, prob = 0.5)
v <- (u + rbinom(n, 1, prob = 0.1)) %% 2
w <- (v + rbinom(n, 1, prob = 0.1)) %% 2
x <- u + 1
y <- v + 1
z <- w + 1
Q(x) + Q(y) + Q(z)
Q(x) + Q(cbind(y, z))
Q(cbind(x, y)) + Q(cbind(y, z)) - Q(cbind(y, z)) - Q(y)
```

[*Hint*] Since these are negative log scores, choose the BN that minimizes the value.

84. In the proof of Proposition 28, prove that if X and Y are independent, then $K_n = 0$ holds regardless of the values of k_X, k_Y, k_{XY}. Moreover, show that (8.5) holds in this case. *Hint:* Use $\alpha\beta - 1 - (\alpha - 1) - (\beta - 1) = (\alpha - 1)(\beta - 1)$.
85. Explain why the check marks in row 2 and below of Table 8.1 appear in those locations.
86. Analogous to (8.8), derive how to compute the eight scores for the MNs in Fig. 3.7.

87. Structure learning for BNs without assuming a variable order:

(a) For the first phase, given data **x** (an $n \times p$ matrix) and p, the following program finds AIC-minimizing parent sets. Line 10 checks whether $x \notin S$. The loop starting at line 12 removes each $Y \in S$ in turn to compare $R(X \mid S \setminus \{Y\})$. After the loop, it compares to $Q(X \mid S)$. This is a double loop over `m` and `i`. The parent set index is stored in `y[m,i]` and its score in `z[m,i]`.

```
1  empty <- matrix(nrow = n, ncol = 0)
2  L <- 2^p
3  y <- array(-1, dim = c(p, L))
4  z <- array(-1, dim = c(p, L))
5  for (m in 1:p) {
6    y[m, 1] <- 1
7    z[m, 1] <- AIC.value(empty, x[, m])
8    for (i in 2:L) {
9      r <- decimal.to.binary(i - 1)
10     if (r[m] == 0) {
11       z[m, i] <- Inf
12       for (j in 1:p) if (r[j] == 1) {
13         s <- r
14         s[j] <- 0
15         k <- binary.to.decimal(s) + 1
16         if (z[m, k] < z[m, i]) {
17           z[m, i] <- z[m, k]
18           y[m, i] <- y[m, k]
19         }
20       }
21       set <- decimal.to.set(i - 1)
22       aic <- AIC.value(as.matrix(x[, set]), x[, m])
23       if (aic < z[m, i]) {
24         z[m, i] <- aic
25         y[m, i] <- i
26       }
27     }
28   }
29 }
```

After understanding the above, run:

```
## Data generation: 4-variable synthetic data
n <- 500    # sample size
p <- 4      # number of variables
x <- matrix(nrow = n, ncol = p)
x[,1] <- rnorm(n)                          # X1: standard normal
x[,2] <- rnorm(n)                          # X2: standard normal
x[,3] <- x[,1] + x[,2] + rnorm(n)          # X3: depends on X1, X2
x[,4] <- x[,3] + rnorm(n)                  # X4: depends on X3
```

Assume the following helpers are defined:

```
decimal.to.binary <- function(x) if (x == 0) NULL else c(x %% 2,
    decimal.to.binary(x %/% 2), rep(0, p))
binary.to.decimal <- function(x) if (length(x) == 0) 0 else x[1]
    + 2 * binary.to.decimal(x[-1])
```

(b) For the second phase, the loop starting at line 12 compares $T(S \setminus \{Y\})R(X \mid S\setminus\{Y\})$ across $Y \in S$. After the loop, it compares to $Q(X \mid S)$. This is again a double loop over `m` and `i`. The arrays store the following: `u[i]`, the score, `v[i]`, the parent set, and `w[i]`, the reduced-set index. Line 30 extracts the removed variables. By line 22, the optimal order and parent sets are determined. In line 27, `t` can contain multiple parents (m of them), and directed edges go from each element of `t` to `j`. Line 31 appends edges, and line 32 updates one level upstream via `k <- w[k]`.

```
1  empty <- matrix(nrow = n, ncol = 0)
2  L <- 2^p
3  u <- array(dim = L)
4  v <- array(dim = L)
5  w <- array(dim = L)
6  v[1] <- 1
7  u[1] <- 0
8  w[1] <- 1
9  for (i in 2:L) {
10   r <- decimal.to.binary(i - 1)
11   u[i] <- Inf
12   for (j in 1:p) if (r[j] == 1) {
13     s <- r
14     s[j] <- 0
15     k <- binary.to.decimal(s) + 1
16     if (u[k] + z[j, k] < u[i]) {
17       u[i] <- u[k] + z[j, k]
18       v[i] <- y[j, k]
19       w[i] <- k
20     }
21   }
22 }
23 edges <- matrix(nrow = 0, ncol = 2)
24 k <- L
25 r <- decimal.to.set(k - 1)
26 while (w[k] > 1) {
27   t <- decimal.to.set(v[k] - 1)
28   m <- length(t)
29   s <- decimal.to.set(w[k] - 1)
30   j <- setdiff(r, s)
31   edges <- rbind(edges, cbind(t, rep(j, m)))
32   k <- w[k]
33   r <- s
34 }
35 edges
```

After understanding the above, run:

```
library(igraph)
g <- graph_from_edgelist(edges, directed = TRUE)
V(g)$label <- 1:p
plot(g)
```

88. For order-free BN structure learning with three variables X, Y, Z, enumerate all expressions of $R(\cdot \mid \cdot)$ and $T(\cdot)$. [*Hint*] There are 12 expressions for R and 7 for T.
89. Prove (8.10).
90. Prove (8.11) and (8.12).
91. Show that the sequences $\{a_n\}$ and $\{b_n\}$ are monotonically nonincreasing. Also prove that equality does not hold in (8.14) and (8.17) for any $n \geq 1$.
92. Replace the function Q in Problem 83 with the BDeu ($\delta = 1$) score and run the computations.
93. In Example 67, verify the inequalities (8.18) and (8.19) for cases (a) and (b). Explain why they correspond to the inequalities between $Q_{X|Y}$ and $Q_{X|YZ}$ under Jeffreys' prior and under BDeu ($\delta = 1$), respectively. Perform the same check for case (b).
94. Suppose we observe n independent trials of (X, Y), where the marginal and joint frequencies are $c_X(x)$, $c_Y(y)$, and $c_{XY}(x, y)$ for $x \in A$, $y \in B$. We estimate the mutual information by

$$I_n(x^n, y^n) := \sum_{x \in A} \sum_{y \in B} \frac{c_{XY}(x, y)}{n} \log \frac{\dfrac{c_{XY}(x, y)}{n}}{\dfrac{c_X(x)}{n} \cdot \dfrac{c_Y(y)}{n}}. \tag{8.31}$$

The code below uses `multi` to sample and `I.n` to compute the estimate. Fill in the blanks and run the procedure.

```
# draw n samples from a multinomial with category probs 'prob'
multi <- function(n, prob) {
  x <- runif(n)
  y <- array(dim = n)
  m <- length(prob)
  for (i in 1:n) {
    for (j in 1:m) {
      if (x[i] < prob[j]) {
        y[i] <- j
        break
      }
      # fill here: subtract prob[j] from x[i] (i.e., shift threshold)
      # x[i] <- x[i] - prob[j]
    }
  }
  return(y)
}
```

```
# mutual information under independence
r <- 100
S <- NULL
for (i in 1:r) {
  x <- multi(100, c(1/2, 1/4, 1/4))
  y <- multi(100, c(1/2, 1/2))
  S <- c(S, I.n(x, y))
}
# mutual information under dependence
T <- NULL
for (i in 1:r) {
  x <- multi(100, c(1/2, 1/4, 1/4))
  y <- (x + multi(100, c(8/10, 1/10, 1/10)) - 1) %% 2 + 1
  T <- c(T, I.n(x, y))
}
# boxplot (S: independent, T: dependent)
boxplot(S, T, names = c("independent", "dependent"),
        main = "Estimated mutual information")
```

95. Using the Asia dataset, apply `I.n`, `J.n`, and `kruskal` to construct a forest (or a tree).
96. Prove (8.22). Then verify on both the `alarm` and `asia` datasets that the Bayesian Chow–Liu algorithm based on (8.21) and on (8.22) yields the same forest.
97. For the Boston dataset, use the $p = 11$ continuous variables (indices 1, 3, 5, 6, 7, 8, 10, 11, 12, 13, 14) and construct a forest using both `I.n` and `J.n` as in the end of Chap. 7.
98. For X, Y Gaussian, one can define J_n analogously to the discrete case so that $J_n \to I(X, Y)$ as $n \to \infty$ and $J_n = 0$ if and only if $X \perp Y$. Below is `I.n` corresponding to the discrete case. Similarly, implement `J.n` for the Gaussian setting, and with $n = 100$, generate random data under independence and dependence to examine whether independence is detected.

```
I.n <- function(x, y) {
  tab <- table(x, y)
  n <- sum(tab)
  px <- rowSums(tab) / n
  py <- colSums(tab) / n
  pxy <- tab / n
  sum <- 0
  for (i in 1:nrow(tab)) {
    for (j in 1:ncol(tab)) {
      if (pxy[i, j] > 0) {
        sum <- sum + pxy[i, j] * log(pxy[i, j] / (px[i] * py[j]))
      }
    }
  }
  return(sum)
}
```

99. Explain the steps from (8.27) to (8.28).
100. When building a forest using J_n, the order of edge insertions can differ from that using I_n. Show that if all variables are binary, the ordering is the same for both I_n and J_n, and explain why.

Bibliography

1. T.W. Anderson, *An Introduction to Multivariate Statistical Analysis*, in Wiley Series in Probability and Statistics, 3rd edn. (Wiley-Interscience, Hoboken, 2003)
2. S. Bochner, *Lectures on Fourier Integrals* (Princeton University Press, Princeton, 1959). Originally published in 1932
3. W. Buntine, Theory refinement on bayesian networks, in *Proceedings of the Seventh Conference on Uncertainty in Artificial Intelligence (UAI)* (Morgan Kaufmann, Burlington, 1991), pp. 52–60
4. C.K. Chow, C.N. Liu, Approximating discrete probability distributions with dependence trees. IEEE Trans. Inf. Theory **14**(3), 462–467 (1968)
5. G. Darmois, Analyse générale des liaisons stochastiques: étude particulière de l'analyse factorielle linéaire. Revue de l'Institut International de Statistique **21**(1/3), 2–8 (1953)
6. J.J. Daudin, Partial association measures and an application to qualitative regression. Biometrika **67**(3), 581–590 (1980)
7. J.K. Ghosh, M. Delampady, T. Samanta, An introduction to Bayesian analysis: theory and methods, in *Springer Texts in Statistics* (Springer, New York, 2006)
8. A. Gretton, K. Fukumizu, C.H. Teo, L. Song, B. Schölkopf, A.J. Smola, A kernel statistical test of independence, in *Advances in Neural Information Processing Systems 20 (NeurIPS 2007)* (Curran Associates, Inc., New York, 2008), pp. 585–592
9. A. Hyvärinen, E. Oja, Independent component analysis: algorithms and applications. Neural Netw. **13**(4-5), 411–430 (2000)
10. M. Kalisch, M. Mächler, D. Colombo, M.H. Maathuis, P. Bühlmann, Causal inference using graphical models with the R package pcalg. J. Stat. Softw. **47**(11), 1–26 (2012)
11. Y. Kano, S. Shimizu, Causal inference using nonnormality, in *Proceedings of the International Symposium on Science of Modeling: The 30th Anniversary of the Information Criterion*, Tokyo, Japan (2003)
12. D. Kelker, Normal distribution: characterizations with applications, in Lecture Notes in Statistics, vol. 100 (Springer, New York, 1995). Revised October 29, 2008
13. J.B. Kruskal, On the shortest spanning subtree of a graph and the traveling salesman problem. Proc. Am. Math. Soc. **7**(1), 48–50 (1956)
14. S.L. Lauritzen, *Graphical Models* (Oxford University, Oxford, 1996)
15. S.L. Lauritzen, D.J. Spiegelhalter, Local computations with probabilities on graphical structures and their application to expert systems (with discussion). J. R. Stat. Soc. Ser. B (Stat Methodol.) **50**(2), 157–224 (1988)
16. Y.V. Linnik, *Decomposition of Probability Distributions* (Oliver and Boyd, Edinburgh, 1968). Translated from the Russian edition (1962)

J. Suzuki, *Graphical Models and Causal Discovery with R*,
https://doi.org/10.1007/978-981-95-4267-3

17. J. Marcinkiewicz, Sur une propriété de la loi de gauss. Mathematicae **9**, 193–196 (1939)
18. R.J. Muirhead, Aspects of multivariate statistical theory, in *Wiley Series in Probability and Mathematical Statistics* (Wiley, New York, 1982)
19. K.P. Murphy, Machine learning: a probabilistic perspective, in *Adaptive Computation and Machine Learning series* (MIT Press, Cambridge, MA, 2012)
20. J. Pearl, *Probabilistic Reasoning in Intelligent Systems: Networks of Plausible Inference* (Morgan Kaufmann, San Mateo, CA, 1988)
21. R.C. Prim, Shortest connection networks and some generalizations. Bell Syst. Tech. J. **36**(6), 1389–1401 (1957)
22. M. Scutari, Learning bayesian networks with the bnlearn r package. J. Stat. Softw. **35**(3), 1–22 (2010)
23. S. Shimizu, P.O. Hoyer, A. Hyvärinen, A. Kerminen, A linear non-gaussian acyclic model for causal discovery. J. Mach. Learn. Res. **7**, 2003–2030 (2006)
24. S. Shimizu, T. Inazumi, Y. Sogawa, A. Hyvärinen, Y. Kawahara, T. Washio, P.O. Hoyer, K.A. Bollen, Directlingam: a direct method for learning a linear non-gaussian structural equation model. J. Mach. Learn. Res. **12**, 1225–1248 (2011)
25. T. Silander, P. Myllymäki, A simple approach for finding the globally optimal Bayesian network structure, in *Proceedings of the 28th Conference on Uncertainty in Artificial Intelligence (UAI)* (2012), pp. 476–485
26. A. Singh, A.W. Moore, Finding optimal Bayesian networks by dynamic programming. Technical Report CMU-CALD-05-106, Center for Automated Learning and Discovery (Carnegie Mellon University, Pittsburgh, 2005)
27. V.P. Skitovich, On a property of the normal distribution. Doklady Akademii Nauk SSSR **89**, 217–219 (1953)
28. P. Spirtes, C. Glymour, R. Scheines, Causation, prediction, and search, 2nd edn. (MIT Press, New York, 2000)
29. E.V. Strobl, K. Zhang, S. Visweswaran, Approximate kernel-based conditional independence tests for fast non-parametric causal discovery. J. Causal Infer. **7**(1), 20180017 (2019)
30. J. Suzuki, Learning bayesian belief networks based on the mdl principle: an efficient algorithm using branch and bound technique, in *Proceedings of the Ninth Conference on Uncertainty in Artificial Intelligence* (Morgan Kaufmann, Washington, D.C., 1993), pp. 266–273
31. J. Suzuki, *Introduction to Bayesian Networks: Fundamentals of Probabilistic Knowledge Processing* (Baifukan, Tokyo, 2009)
32. J. Suzuki, The Bayesian chow-liu algorithm, in *Proceedings of the 6th European Workshop on Probabilistic Graphical Models (PGM 2012)* (Granada, Spain, 2012), pp. 315–322
33. J. Suzuki, A theoretical analysis of the bdeu scores in bayesian network structure learning. Behaviormetrika **44**(1), 97–116 (2017)
34. J. Suzuki, *Statistical Learning with Math and R: 100 Exercises for Building Logic* (Springer, Singapore, 2020)
35. J. Suzuki, *Statistical Learning with Math and Python: 100 Exercises for Building Logic* (Springer, Singapore, 2021)
36. J. Suzuki, *Kernel Methods for Machine Learning with Math and Python: 100 Exercises for Building Logic* (Springer, Singapore, 2022)
37. J. Suzuki, *Kernel Methods for Machine Learning with Math and R: 100 Exercises for Building Logic* (Springer, Singapore, 2022)
38. J. Suzuki, J. Kawahara, Branch and bound for regular bayesian network structure learning, in *Proceedings of the 33rd Conference on Uncertainty in Artificial Intelligence (UAI)* (AUAI Press, Washington, 2017)
39. J. Suzuki, T. Yang, Generalization of LiNGAM that allows confounding, in *2024 IEEE International Symposium on Information Theory (ISIT)* (IEEE, New York, 2024), pp. 3540–3545
40. T. Verma, J. Pearl, Causal networks: Semantics and expressiveness, in *Proceedings of the Fourth Annual Conference on Uncertainty in Artificial Intelligence (UAI-86)* (AUAI Press, Mountain View, CA, 1986), pp. 69–76

41. Y.S. Wang, M. Drton, Causal discovery with unobserved confounding and non-gaussian data. J. Mach. Learn. Res. **24**(79), 1–62 (2023)
42. S. Watanabe, *Theory and Methods of Bayesian Statistics* (Corona Publishing, Tokyo, 2012)
43. K. Zhang, J. Peters, D. Janzing, B. Schölkopf, Kernel-based conditional independence test and application in causal discovery, in *Proceedings of the Twenty-Seventh Conference on Uncertainty in Artificial Intelligence (UAI 2011)* (AUAI Press, Mountain View, CA, 2011), pp. 804–813

Index

J. Suzuki, *Graphical Models and Causal Discovery with R*,
https://doi.org/10.1007/978-981-95-4267-3

MIX
Papier aus verantwortungsvollen Quellen
Paper from responsible sources
FSC® C105338

If you have any concerns about our products, you can contact us on
ProductSafety@springernature.com

In case Publisher is established outside the EU, the EU authorized representative is:
Springer Nature Customer Service Center GmbH
Europaplatz 3, 69115 Heidelberg, Germany

Printed by Libri Plureos GmbH
in Hamburg, Germany